Edinburgh, 3 March 2019

Mastering Academic Writing in the Sciences

To Ghina,

With deep admiration of a wonderful speaker and writer

Marialuisa

Mastering Academic Writing in the Sciences

A Step-by-Step Guide

Marialuisa Aliotta

CRC Press
Taylor & Francis Group
Boca Raton London New York

CRC Press is an imprint of the
Taylor & Francis Group, an **informa** business

CRC Press
Taylor & Francis Group
6000 Broken Sound Parkway NW, Suite 300
Boca Raton, FL 33487-2742

© 2018 by Taylor & Francis Group, LLC
CRC Press is an imprint of Taylor & Francis Group, an Informa business

No claim to original U.S. Government works

Printed on acid-free paper
Version Date: 20180306

International Standard Book Number-13: 978-1-138-74288-8 (Hardback)
International Standard Book Number-13: 978-1-4987-0147-1 (Paperback)

This book contains information obtained from authentic and highly regarded sources. Reasonable efforts have been made to publish reliable data and information, but the author and publisher cannot assume responsibility for the validity of all materials or the consequences of their use. The authors and publishers have attempted to trace the copyright holders of all material reproduced in this publication and apologize to copyright holders if permission to publish in this form has not been obtained. If any copyright material has not been acknowledged please write and let us know so we may rectify in any future reprint.

Except as permitted under U.S. Copyright Law, no part of this book may be reprinted, reproduced, transmitted, or utilized in any form by any electronic, mechanical, or other means, now known or hereafter invented, including photocopying, microfilming, and recording, or in any information storage or retrieval system, without written permission from the publishers.

For permission to photocopy or use material electronically from this work, please access www.copyright.com (http://www.copyright.com/) or contact the Copyright Clearance Center, Inc. (CCC), 222 Rosewood Drive, Danvers, MA 01923, 978-750-8400. CCC is a not-for-profit organization that provides licenses and registration for a variety of users. For organizations that have been granted a photocopy license by the CCC, a separate system of payment has been arranged.

Trademark Notice: Product or corporate names may be trademarks or registered trademarks, and are used only for identification and explanation without intent to infringe.

Library of Congress Cataloging-in-Publication Data

Names: Aliotta, Marialuisa, author.
Title: Mastering academic writing in the sciences : a step-by-step guide / Marialuisa Aliotta.
Other titles: Academic writing in the sciences
Description: Boca Raton, Florida : CRC Press, [2019] | Includes bibliographical references and index.
Identifiers: LCCN 2018001536| ISBN 9781138742888 (hardback : alk. paper) | ISBN 9781498701471 (pbk. : alk. paper) | ISBN 9781351002141 (master ebook) | ISBN 9781351002127 (epub) | ISBN 9781351002134 (ebook pdf) | ISBN 9781351002110 (ebook kindle)
Subjects: LCSH: Communication in science--Handbooks, manuals, etc. | Technical writing--Handbooks, manuals, etc. | Academic writing--Handbooks, manuals, etc. | Language arts--Correlation with content subjects--Handbooks, manuals, etc.
Classification: LCC Q223 .A37 2019 | DDC 808.06/65--dc23
LC record available at https://lccn.loc.gov/2018001536

Visit the Taylor & Francis Web site at
http://www.taylorandfrancis.com

and the CRC Press Web site at
http://www.crcpress.com

To Lorenzo,
the joy of my life.

Contents

List of Figures — xiii

List of Tables — xv

Preface — xvii

Introduction — xxi

SECTION I FOUNDATIONS

CHAPTER 1 ▪ Before You Begin — 3
- 1.1 WHAT IS ACADEMIC WRITING? — 3
- 1.2 WRITING AS A CRAFT — 5
- 1.3 GETTING STARTED — 6
- 1.4 THE WRITER'S MINDSET — 6
- 1.5 STRATEGIES TO AVOID PROCRASTINATION — 7
  - 1.5.1 Managing time — 8
  - 1.5.2 Managing distractions — 9
  - 1.5.3 Staying on track: Weekly check-ups — 9
- 1.6 YOUR WRITING SKILLS: SELF-ASSESSMENT CHECKLIST — 12

SECTION II THE WRITING PROCESS

CHAPTER 2 ▪ Pre-Writing Step — 19
- 2.1 GATHERING AND ORGANISING INFORMATION — 20
- 2.2 THE THREE OBJECTIVES OF READING — 21
  - 2.2.1 Capturing content: Taking notes while reading — 21
    - 2.2.1.1 The paper annotation tool — 22
  - 2.2.2 Capturing structure: Developing a template — 24
  - 2.2.3 Capturing style: Learning from the masters — 25

	2.3	THE LITERATURE REVIEW	26
		2.3.1 What it is and why it matters	26
		2.3.2 The literature review as a process	27
		2.3.3 Critical voice	27
		2.3.4 The literature review as a product	31
		2.3.5 How to write a literature review	32
		2.3.6 Literature review matrix	33

CHAPTER 3 ▪ The Drafting Step — 39

	3.1	WHO IS YOUR AUDIENCE?	39
	3.2	TALKING ABOUT YOUR RESEARCH	41
	3.3	GETTING THE STRUCTURE RIGHT	43
	3.4	MIND MAPS	43
	3.5	CORE DUMP	47

CHAPTER 4 ▪ The Revising Step — 51

	4.1	THE TRIAGE APPROACH	52
	4.2	COMMON PROBLEMS	52
		4.2.1 Faulty organisation	52
		4.2.2 Lack of clarity	53
		4.2.3 Inappropriate usage of language	55
		4.2.4 Poor grammar	56
	4.3	TIPS FOR A BETTER STRUCTURE	56
	4.4	PARAGRAPHS AS BUILDING BLOCKS	58
	4.5	REVERSE OUTLINING	59
	4.6	LINKING PARAGRAPHS TOGETHER	60
	4.7	PARALLEL STRUCTURE	61
	4.8	FEEDBACK: WHEN, WHAT, AND WHOM TO ASK	63

CHAPTER 5 ▪ The Editing Step — 67

	5.1	HOW GOOD IS YOUR WRITING?	68
	5.2	SCIENTIFIC STYLE IN ACADEMIC WRITING	69
	5.3	PEARLS OF WISDOM: ADVICE FOR A BETTER STYLE	72
		5.3.1 Verbs in action	73
		5.3.2 Verbs in disguise	74
		5.3.3 Active or passive?	75

	5.3.4 *I* or *we*? Personal pronouns in scientific writing	76
5.4	DE-CLUTTER YOUR TEXT	77
	5.4.1 Awkward phrases and waste words	78
	5.4.2 Transition words	78
	5.4.3 Redundant information	79
	5.4.4 Negative statements	79

CHAPTER	6 ▪ The Proofreading Step	83
6.1	WHEN DETAILS MATTER	84
6.2	COMMON GRAMMAR MISTAKES	84
	6.2.1 Homophones	85
	6.2.2 Subject-verb concordance	85
	6.2.3 Singular or plural?	85
	6.2.4 Dangling participle	86
	6.2.5 Ambiguous referencing	87
6.3	PUNCTUATION MARKS OFTEN MISUSED	88
	6.3.1 Comma	88
	6.3.2 Semi-colon	90
	6.3.3 Colon	90
	6.3.4 Hyphen	91
	6.3.5 Apostrophe	91
6.4	SPELLING CHECKS	92
6.5	CITATIONS AND BIBLIOGRAPHIES	93
	6.5.1 What to cite, where, and how	94
	6.5.2 Reference formats	94
6.6	PROOFREADING CHECKLIST	96

CHAPTER	7 ▪ The Technical Stuff	99
7.1	TITLES	99
7.2	TABLE OF CONTENTS	101
7.3	FIGURES AND TABLES	102
	7.3.1 What goes in a figure	102
	7.3.2 Figure captions	103
	7.3.3 Examples of poorly prepared figures	104
	7.3.4 What goes in a table	107
	7.3.5 Table titles	107

7.4	EQUATIONS AND SYMBOLS	108
7.5	REPORTING EXPERIMENTAL RESULTS	109
7.6	APPENDICES	109
7.7	GLOSSARY AND LISTS OF ACRONYMS	110
7.8	ACKNOWLEDGMENTS	111

Chapter 8 ▪ Worked-out Examples — 115

8.1	ENERGY CONSUMPTION IN DATA CENTERS	115
8.2	COLORECTAL CANCER	120
8.3	BLAINVILLE'S BEAKED WHALES	127

Section III Supporting Material

Chapter 9 ▪ Section Templates — 143

9.1	INTRODUCTION		144
	9.1.1	The purpose	144
	9.1.2	Building a template: A worked-out example	145
	9.1.3	Introduction: A template	147
9.2	METHODS		149
	9.2.1	The purpose	149
	9.2.2	Methods: A template	150
9.3	DATA ANALYSIS AND RESULTS		151
	9.3.1	The purpose	151
	9.3.2	Data analysis and results: A template	152
9.4	DISCUSSION AND CONCLUSIONS		152
	9.4.1	The purpose	152
	9.4.2	Discussion and conclusions: A template	154
9.5	ABSTRACT		155
	9.5.1	The purpose	155
	9.5.2	Abstract: A template	155

Chapter 10 ▪ Elements of English Grammar — 159

10.1	BASIC TERMS AND DEFINITIONS		160
	10.1.1	Clauses and sentences	160
	10.1.2	Subject	160
	10.1.3	Object	160
	10.1.4	Nouns	161

	10.1.5	Pronouns	162
	10.1.6	Adjectives and adverbs	162
	10.1.7	Prepositions and conjunctions	162
	10.1.8	Verbs	163
	10.1.9	Phrasal verbs	164
	10.1.10	Verb tenses	165
	10.1.11	Infinitives, participles, and gerunds	167
	10.1.12	Auxiliary and modal verbs	168
10.2	SIMILAR MEANING, DIFFERENT SPELLING		168
	10.2.1	*Due to* or *owing to*?	168
	10.2.2	*That* or *which*?	169
	10.2.3	*Fewer* or *less*?	169
	10.2.4	*Who* or *whom*? and other personal pronouns	170
10.3	SIMILAR SPELLING, DIFFERENT MEANING		170

References   173

Index   175

List of Figures

2.1 Example of a literature matrix showing key research aspects (columns) and their corresponding outcomes (rows) as reported in various papers. A literature matrix can simplify the task of critically comparing and contrasting existing literature.   34

3.1 Example of a mind map showing a central topic and its various branches with related items connected by lines.   44

3.2 Same map as the one in Figure 3.1 but with all its elements arranged by association.   45

3.3 Example of a structured layout from the mind maps of Figures 3.1 and 3.2. Producing a structured layout also gives you a good idea of how much text to write in each chapter (section) based on the estimated total length of your thesis (paper).   46

5.1 Sample outcome of the Writer's Test for an early draft excerpt.   69

7.1 Excerpt from the Table of Contents of a PhD thesis on the study of a nuclear reaction in Classical Novae. Note how descriptive titles provide a good indication of what the reader can expect to find in each section and sub-section.   102

7.2 Example of a poorly prepared figure from a student's internal report showing a number of issues as discussed in the text.   105

7.3 Example of a poorly prepared figure from a student's internal report showing a number of issues as discussed in the text.   106

7.4 Example of a complex, yet well laid-out table. The table title provides enough information to allow the reader to understand what is presented in the table.   107

List of Tables

4.1	Overview table on *when*, *what* and *whom* to ask for feedback on your writing.	64
5.1	Useful verbs and adverbs indicating various degrees of certainty.	72
5.2	Examples of strong verbs often found in academic writing.	74
5.3	Abstract nouns often used in scientific writing. For better style, replace the nouns with the verbs they disguise whenever possible.	76
6.1	Singular and plural forms of commonly used words of Latin and Greek origin. Make sure they concord properly with the verb used.	86
6.2	Apostrophe's functions.	92
9.1	INTRODUCTION: A template (Adapted from *Science Research Writing for Non-Native Speakers of English*, by Glasman-Deal [8])	148
9.2	METHODS: A template (Adapted from *Science Research Writing for Non-Native Speakers of English*, by Glasman-Deal [8])	150
9.3	DATA ANALYSIS AND RESULTS: A template (Adapted from *Science Research Writing for Non-Native Speakers of English*, by Glasman-Deal [8])	153
9.4	DISCUSSION AND CONCLUSIONS: A template (Adapted from *Science Research Writing for Non-Native Speakers of English*, by Glasman-Deal [8])	154
9.5	ABSTRACT: A template (Adapted from *Science Research Writing for Non-Native Speakers of English*, by Glasman-Deal [8])	156
10.1	Verb tenses (present, past, and future) conjugated in their various forms (simple, continuous, perfect, and perfect continuous).	165
10.2	Commonly confused words often found in scientific writing.	171

Preface

Publishing is a crucial component of doing research. One might argue that unless we are able to communicate effectively the outcome and relevance of the research we do, little progress can be achieved. Yet, writing does not generally come easily to most of us. Students often struggle with learning the crafts of the trade and their supervisors, generally overloaded with other demands on their time, may not be able to provide the necessary guidance and support.

This book builds on the experience I have gained from running three-day workshops for physics students in Scotland (UK) for almost a decade. The success of those workshops convinced me about the need to provide clear instructions on how to approach writing in a scientific subject.

Unlike other types of academic writing, scientific writing has a clearly identifiable structure and obeys commonly followed "rules" that other researchers will easily recognise and understand. In some ways, then, scientific writing is *easier* than people may think. Yet, many students and other inexperienced researchers often struggle to master the skills required for good scientific writing and often end up feeling overwhelmed, stressed and anxious about their ability to produce a cohesive PhD thesis or research paper as a comprehensive account of years of research.

This book has been designed with the intention of easing that overwhelming feeling by providing a clear step-by-step framework that anyone can put into practice and that can be equally applied to a thesis, a dissertation or a research paper.

Who can benefit from using this book

Even though this book is primarily intended for PhD and Masters students, in my experience any academic (whether young or established) who struggles with scientific writing can benefit from reading it and adopting its step-by-step framework as a guiding strategy to their writing.

While different scientific disciplines may pose different challenges and follow conventions slightly different from those presented in this book, the approach to writing can (and ideally should) be the same: it progresses through a clear sequence of steps, each one tackling a specific aspect of the writing process. My purpose (and hope) in writing this book is to offer you a roadmap to follow so that your writing may become a productive and enjoyable activity.

How to use this book

The book is organised into three sections. **Section I** provides some **foundations** to help you prepare for the work ahead. It offers a few simple strategies to support you with overcoming procrastination and avoiding distractions.

Section II focuses entirely on **the writing process** through five key steps, namely: pre-writing, drafting, revising, editing, and proofreading. Each step comes with an explanation of what is involved and contains specific tasks, in the **Exercises** section, designed for you to practice and consolidate the ideas presented. In the same way as you would not be able to speak a foreign language if you did not practise it, you will not be able to improve your writing just by *reading* this book. So, take the time to go through as many tasks as possible.

My advice would be for you to quickly skim through **Section II** in order to get an overview of the entire process involved with writing a thesis or research paper. Then, once you are ready to begin writing, focus on one specific thesis chapter (or section, if you are writing a research paper) and complete all the steps from pre-writing to proofreading. Only move to a new thesis chapter or paper section once you have gone through the full framework. This will allow you to familiarise yourself with the amount of work (and time!) involved in producing a well crafted (and hopefully final) chapter or section.

Finally, **Section III** consists of two chapters of **supporting material**: Section Templates (Chapter 9) and Elements of English Grammar (Chapter 10). Chapter 9 describes the meaning and purpose of specific sections (Introduction, Methods, Data Analysis and Results, Discussion and Conclusions, Abstract) typically encountered in scientific writing. It also shows how to create templates for those sections that you can use in your own writing. So, make sure you read the relevant section in this chapter before you attempt to write any of the Introduction, Methods, Data Analysis and Results, Discussion and Conclusions, Abstract section or chapter in your own paper or thesis.

Chapter 10 is essentially meant for reference use. Throughout the book, I refer to various grammatical elements to describe different parts of a sentence. If you are not familiar with these, please make sure you read Chapter 10 first.

Additional key features of the book include worksheets (such as the Weekly Check-up and the Paper Annotation Tool), as well as a full chapter (Chapter 8) of **Worked-out examples** to illustrate how early drafts can be transformed into more polished revisions by identifying and tackling, in turn, the most pressing issues of the prose.

A word of caution

If you have ever tried learning to play a sport or an instrument on your own and have later decided to take proper instruction, you may have experienced a sense of lack of progress at first or, worse, a sense of *going backwards* almost as if you were making more progress on your own rather than with your instructor. This happens because as soon as we take proper lessons, we discover that we first need to *unlearn* all the bad habits that we have picked up as a result of self-teaching.

Writing is no different. So, when you first go through the framework presented in this book, you may feel that you are making less progress than before, or that progress is slower than you were used to before applying the step-by-step approach from this book. If so, do not get discouraged, but trust the process instead. With practising the steps in the order they are intended, you will soon realise that it takes less and less effort to go from an early draft to its final polished version. Eventually, the tools and techniques presented here will feel second nature to you and you will start producing better initial drafts as a result. Just trust the process and re-assess your writing abilities after you have completed it a few times.

Sources

I have consulted many books on writing (both academic and non) when preparing the material for my workshops and for this book. While I have tried to give proper reference to all my sources throughout the book and listed additional ones in the **Further reading** section of most chapters, I may have missed some. For this I am solely responsible and I can only state here that any omission was not intentional.

Acknowledgements

I have a debt of gratitude to all the students in my workshops, who have provided much insight into their struggles and challenges. I am particularly grateful to those who have given me permission to re-use their material as examples for the book, and to my colleagues, John Loveday and Malcolm McMahon, for the abstracts in Chapter 1.

My heartfelt thanks also go to Francesca McGowan, Acquiring Editor at Taylor & Francis, who first believed in this project and enthusiastically supported me through its early stages. Francesca and her Editorial Assistants, Emily Wells and Rebecca Davies, have been superb and quite simply the best editors I could have hoped to work with. To them goes my gratitude for their patience and understanding during times of delay due to my busy schedule and for their unwavering support and encouragement throughout the process. I also wish to thank all the members of the production team, especially Robin Lloyd-Starkes and Shashi Kumar, and the design team for translating my original idea into what I believe is a cheerful and captivating front cover.

Finally, I am grateful to Carlo Bruno, Tom Davinson, Gary Docherty, Maria Francesca Marzioni, and Aurora Tumino for much welcome feedback on early drafts of some chapters.

And last but not least, a huge thank you to my family for always believing in me even when I do not, and to Malcolm, for his presence and patience, for his support in every new project I undertake, for the wonderful basecamp we have created for each other, and for being a wonderful dad to our son.

Introduction

I still remember when, as an undergraduate, I handed in a draft of my dissertation's first chapter to my supervisor, Prof. Aldo Zappalà. I was feeling very proud of what I had done and was looking forward to rave feedback. So, a few days later, when I went to see him to discuss my draft, I did so in a state of anticipation and even though I did not exactly expect a pat on the shoulder, I was confident I would get a big, resounding "well done!".

Boy, was I wrong! What I got back was a printed out copy of my chapter completely scribbled over in red pen and full of question marks, corrections, and entire paragraphs crossed out. It was painful. Everything seemed to be wrong: the style, the structure, the flow of arguments. I remember leaving my supervisor's office with a sense of despair and the certainty that I would never make it to graduation. To make things worse, I was left with a sense of disbelief. How was it possible for my writing to be so poor!? After all, I had always been a very conscientious student; one who used to take notes regularly while studying; one who knew where to look for every paper I had read; one who had always been strongly motivated to do well and to succeed. I was incredulous.

Luckily, my supervisor had seen plenty of similar situations throughout his career and did not seem discouraged in the least. He said my chapter was "a good start". "What?! You must be crazy – I thought – A good start? How come then that there is so much red ink all over the pages?" – I wondered. As if he could read my thoughts, he placidly went on to add that at least there was some *content* to work with. Surely, there was still plenty "to improve on the structure and on the style" – he said – "but at least the content *is* there". He had seen worse, he added. And I gasped!

The months that followed were not a bit less painful than our first meeting. On countless evenings I went back home feeling intensely discouraged and desperate at the looming prospect of never managing to write something that would stand the test of my supervisor's standards and expectations of me. Yet, week after week, by cutting (literally, with scissors!) and pasting (literally, with glue!) bits and pieces of my own writing, by constantly revising and editing, I finally got the hang of it. I started to see what I was doing wrong; what I needed to change; how I was supposed to express concepts, findings, and methods; and slowly, page after page, I learned the process of writing a dissertation.

With hindsight, I am actually very pleased that this is the way things went for me, because I learned a lot! I learned how to structure my writing properly, how to make sure the flow of text was logical, how to separate methods from results and how to adopt an appropriate scientific style. As it turned out, this learning process was a blessing for my PhD, because when it came to writing up my thesis I knew exactly how to do so and did not waste any time in trials and errors.

Interestingly, I now realise that if I had been able to write well straightaway I might have never bothered to find out what I was doing right, how, and why. Having discovered that for myself, I can help others to do the same. But, perhaps, the most valuable lesson I learned out of this experience was that academic writing is a craft. And like any craft, it can be learned. I am indebted to my supervisor for taking the time to teach me much of what I know about academic writing.

Nowadays, when my students hand in a report or a chapter of their thesis, I know that, very likely, they will come back to me a few days later with the same anticipation I had. And they too, will leave my office feeling a similar sense of frustration when realising there is still so much they have to work on. Most of the times, my assessment of their work is the same as that of my supervisor. "At least, the content is there" – I say to them with the most encouraging tone I can muster – "Now let's work on everything else."

In fact, year after year, I have found that PhD students stumble upon the same mistakes not as much in their actual writing (though there are some recurring themes there as well), but more in their attitude to the whole "writing-up" business. Luckily, there are strategies and tips that you can follow to avoid these and similar traps. Sometimes, all you need is someone to show you how to do it.

This book was written with you in mind in the hope that avoiding similar mistakes will save you lots of time and frustration when you eventually will sit down and start writing up! I hope you will find it useful. Just let me know. I would love to hear back from you.

I

FOUNDATIONS

CHAPTER 1

Before You Begin

CONTENTS

1.1	What is academic writing?	3
1.2	Writing as a craft	5
1.3	Getting started	5
1.4	The Writer's Mindset	6
1.5	Strategies to avoid procrastination	7
	1.5.1 Managing time	8
	1.5.2 Managing distractions	9
	1.5.3 Staying on track: Weekly check-ups	9
1.6	Your Writing Skills: Self-Assessment Checklist	12

1.1 WHAT IS ACADEMIC WRITING?

The term *academic writing* refers to a specific type of prose used by scholars within an academic context. Some academic writing is done for the purpose of fulfilling requirements of a university or college, for example through the submission of a PhD thesis or dissertation; some is done as a way to disseminate results or information, for example through research papers, conference proceedings, collaboration reports; and some is done to attract funds, through grants or fellowships applications. How successful you are in each of these tasks ultimately depends on how well you can write in a proper *academic* way.

While many different types of academic writing outputs exist, the defining features of academic writing are broadly the same. In scientific disciplines, they include: a well-defined, recognisable structure; a formal tone (or *register*) free from colloquialisms; a factual perspective, typically centred on objective, often experimental evidence; a clear focus on the research question under study; an accurate choice of words that avoids ambiguity; and an analytical approach (or deductive reasoning) that presents a logical, consequential flow of argument.

Thus, in many respects, academic writing can be regarded as a kind of specialised language used by scholars to convey a body of information about a particular subject. And like any language, it can be learned. This, however, requires an understanding of its basic rules and a willingness to put them into practice.

All too often, inexperienced academic writers assume that their writing should be full of impressive words and technical terms for it to be considered *academic* enough. This type of writing, however, confuses the reader and does not fulfil the primary function of writing, namely that of disseminating knowledge. Excessive use of jargon, or highly technical terms, normally obfuscates the meaning and leaves the reader unclear about the author's intended message.

On the contrary, good academic writing should be *clear*, *concise* and *pleasant* to read. Unfortunately, this is easier said than done. In part, this is because many myths and misconceptions still exist around what makes for good academic writing. In part, it is because academics do not normally receive proper training in academic writing. Somehow, the expectation is that researchers will learn by "osmosis" and pick up their writing skills along the way.

The two examples below help illustrating these points. What do you like or dislike about each example? Which one, if any, attracts your attention and why? Can you identify features or 'good' and 'bad' writing in either? And finally, what makes for good academic writing in your opinion?

As we shall see in the next chapter, much of our ability to write well stems from our ability to critically assess what we read. Throughout the rest of the book we will then discuss key principles of structure and style that contribute to improving our writing and will present specific tasks involved at different stages of writing.

Abstract for a Colloquium

A HEURISTIC MODEL OF METHANE ENCLATHRATION

A thermo-physical model has been developed for the enclathration of methane in ice based on the Kuhs formulation of the Langmuir isotherm but using new heuristic potentials for both methane and water obtained by the McMahon method.

The model predicts multiple filling of both large and small water cages with quintuple occupancy of the large cages at the highest pressure studied. The model is compared with time-of-flight diffraction studies by Loveday and co-workers and suggests that these studies overlooked the effect of Edgeworth anharmonicity on the measured cage occupancies.

The consequences of the model for important phenomena like the Titanian core overturn event and the Eocene extinction event will be discussed.

> Abstract for a General Interest Seminar
>
> ## NEVIL MASKELYNE:
> ## THE MAN WHO USED A MUNRO TO WEIGH THE EARTH
>
> In the "summer" of 1774, the Astronomer Royal, Nevil Maskelyne, spent 4 months more than 2,000ft up the side of one of Scotland's most well known mountains conducting an experiment, first suggested by Isaac Newton, to determine the mean density of the earth.
>
> Using a 10-foot telescope, and after making more than 300 observations on 43 different stars, he was able to show that the earth had a dense solid core, and was not an "immense vacuity".
>
> In this talk, I will describe Maskelyne's famous experiment, and also a very similar experiment conducted somewhat closer to King's Buildings (Edinburgh, UK). Finally, I will describe a recent attempt to repeat Maskelyne's original experiment using little more than topographic data from the Space Shuttle.

1.2   WRITING AS A CRAFT

Unlike reading, writing is not normally a "linear" activity and it does not normally take place in one setting, unless you are either extremely gifted or very experienced (in which case you probably would not be reading this book!).

When we read a well-written thesis or paper, we are presented with a text whose component parts form a logical sequence, even though the order in which they are presented is rarely the same one in which they were written.

We also often underestimate the amount of work and revision that a given document has undergone. So, it is vital to realise that good academic writing requires careful planning; a clear idea of what to say; an acceptance that first drafts will be far from perfect; and the willingness to spend enough time revising and polishing the text.

Appreciating these aspects can make the whole difference to the way we approach, and feel about, our writing. So in many ways, mastering academic writing is a craft and like with any craft, it requires practice and dedication. There are no right or wrong approaches to writing, but finding the one that works for you is crucial to make it a success.

1.3 GETTING STARTED

If you are a PhD student, you may decide to start writing from an early stage of your doctoral training. This has the advantage of helping generate ideas, keeping motivation, and providing useful practice. The downside is that you may end up with large chunks of writing that will never make it to your final thesis.

Alternatively, you may decide to do all your writing towards the end of your PhD, normally after completing your data analysis. This approach has the advantage that, by then, you will have a clearer idea about what to write and will have spent less time on "wasted" writing. The disadvantage, of course, is that you will have gained less practice and it may take you longer to achieve the level of mastering in your writing.

Whether you decide to write from the beginning or towards the end of your PhD, you also need to decide how to approach the actual act of writing. Some people prefer to establish a daily routine (for example, 30 minutes every day, or 500 words every day); others work better with a goal-centred approach (for example, completing a whole section or chapter; writing a comprehensive literature review) regardless of the time it takes and only move to the next task once a given one has been completed. This is largely a matter of personal choice and you may want to try both approaches to find out what works best for you.

Whatever your strategy, you need to make writing a top priority in your daily routine. This means regarding your writing in the same way as you would regard other important commitments and protecting the time you allocate to it from other external demands. In later chapters, we will talk more about useful strategies to overcome the writer's block and make it easier on you to actually start writing. But first, let us start from an appropriate mindset that will secure steady progress.

1.4 THE WRITER'S MINDSET

Mindset is hugely important to your success as a writer. Most students approach writing with little preparation, if any at all. At some point, normally towards the end of their PhD, they decide (or realise) that it is time to write up and make a conscious effort to sit before their favourite word processor hoping for some inspiration.

Sadly, this approach rarely works; more often it ends up in frustration, and unless it is properly rectified, it may lead to total disengagement. In fact, you may have already experienced the frustration of writing something without really making any progress. Maybe you take on board some feedback from your supervisor, but soon you begin to feel confused, insecure, overwhelmed, and eventually become stressed and anxious about whether you will ever manage to submit your thesis or paper.

A far better strategy is to plan carefully and prepare yourself in the best possible way *before* you begin to write, so that you can experience ease, confidence, and flow – possibly even fun. This strategy, first and foremost, requires *clarity* about what you want to say. And to achieve clarity you need to start with your own mind, not with your word processor! Indeed, this is so important that I have devoted the whole of Chapter 2 (*The Pre-Writing Step*) and part of Chapter 3 (*The Drafting Step*) to the process of getting a clear idea of what to write before committing any words on paper. Only then can you begin to write unless, of course, you get stuck in procrastination.

1.5 STRATEGIES TO AVOID PROCRASTINATION

Procrastination can be defined as a *"self-defeating behaviour pattern"* characterised by *"short-term benefits and long-term costs"* [1].

This reminds me of a PhD student in our department, some years ago. She had performed all her experiments and completed the related data analysis; she had a clear plan of what to write; and she was conscious that time to write up had finally arrived. Yet, she was struggling with making a start (that sounds familiar, I know). One day, she posted a picture on Facebook of her beautifully polished fingernails in red varnish with a comment that read: "I have never ever polished my fingernails in my entire life, but I would do anything as long as it is not writing my thesis!"

To some extent, we all suffer from the tendency to delay or postpone something, especially when the task ahead is daunting, unknown or unpleasant.

Procrastination has many causes: fear of failure, fear of success, lack of interest, lack of motivation, lack of time, excessive perfectionism, lack of clarity about what to do, stress, depression, anxiety... just to name a few. Of course, suggesting ways to tackle any of these root causes would go beyond the purpose of a book like this, but when it comes to writing, some people seem to find comfort in accepting and practising imperfection. Others decide to join a writing group or to keep a writing logbook to assess their progress, as suggested in *Becoming an Academic Writer* [2] by Patricia Goodson. Others settle for an accountability partner or finally resolve to hire a mentor.

Whatever strategy you decide to adopt, you will soon realise that the core element of any anti-procrastination strategy is always to take *some* action. The focus here, though, is not on "taking action" (which in itself is precisely what procrastination prevents you from doing) but rather on "some". It does not matter how big or small your first step is, as long as it is a step in the right direction towards a very specific goal.

One way to help you crystallise that action into something resembling a habit is to try to embed some form of reward in your activity: an evening at the cinema once you complete a chapter; an ice cream for a camera-ready figure; half-an-hour reading your favourite book; a 15-min walk in the park; you get the idea. Anything that gives you joy or a sense of well-being will do, as long as it creates a positive reinforcement and helps you to establish a productive routine.

If everything else fails, consider finding a "writing buddy": taking a commitment with someone else can work wonders to combat our own inactivity. If having company is all you need to help you get going with you writing, by all means find some!

1.5.1 Managing time

Once you have begun to take action, you need to keep momentum. In my experience, the two most critical factors that may prevent you from establishing a successful writing routine are: 1) how well you manage your time, and 2) how effectively you manage distractions.

Time flows at the same rate for all of us[1]: we all have 24 hours a day. So, how come that some people are far more productive than others? The answer is simple: they use their time more effectively. During my school years, I used to attend a ballet class for an hour and a half every afternoon and I would not be home until 6.30pm every evening. As a result, I only had a couple of hours left to do my homework before dinner at 9pm (yes, we eat that late in Sicily!). It was not much time, but I knew it was the only time I had and for two hours flat I would just study without distractions or interruptions.

These days, one of my most productive days or weeks are those that precede my going on holiday. Suddenly, having a finite amount of time to complete any pending job works wonders for me: I become much more focussed and able to concentrate better on the task on hand. By contrast, when I have full days free from meetings, teaching, or other commitments, I tend to waste much of my time in pointless activities and indulging in useless distractions (ok, perhaps I should not disclose this so openly, but you get the point).

Maybe you relate. If so, a powerful technique to establish a successful writing routine (and generally improve your productivity) consists in *framing your time*. This is beautifully illustrated in the Pomodoro Technique [3], a simple yet effective approach to managing your time in a productive way. All you need to do is to set a kitchen timer (typically 25-min chunks is all you need) and work exclusively on a specific task (writing in our case) without interruptions or distractions. If you have never done anything like this, you may feel slightly awkward at first, and you may need to practise a little before you obtain tangible results. However if you keep practicing, you will soon be amazed by how much you can accomplish in little chunks of time!

In my 3-day writing workshops, I make extensive use of the Pomodoro Technique (even though I do not explicitly refer to it as such). After introducing a given topic, I usually give my students 15-20 minutes to carry out a specific task and put things into practice. By the end of the workshop, and much to their surprise, many students report having been able to write a good deal of text, revise it, and polish it to near completion.

---

[1] Unless, of course, you are travelling at the speed of light, which for the moment I am assuming you are not.

Similarly, many of the exercises proposed in this book are framed within a set amount of time. While framing your time may seem mechanical at first, it can provide an excellent starting point to establishing a routine. Eventually, you will not need a timer anymore: you will feel the enjoyment that comes from having achieved something and will naturally want to continue doing so.

1.5.2 Managing distractions

I once heard that it takes about 15 minutes to re-focus our attention to a specific task once we have been distracted from it. Distractions can take the shape of a Facebook message, a phone call, a mail pop-up alert, or a sudden thought about what to cook for dinner. To avoid falling prey to distractions you need to learn how to manage them. Sometimes, closing your mail system or your Facebook page, or even disconnecting the telephone (at least for an amount of time that you feel comfortable with), is all you need to do.

More difficult may be to deal with internal chats and thoughts that constantly bombard our minds. A strategy that I have personally tried and found useful consists in spending a couple of minutes to clear my mind before starting to write. I do so by taking some quick notes about any thoughts and "to do" items that pop up to mind. By writing them down on a piece of paper, I subconsciously tell my brain that I know they are there and I will take care of them as soon as I can, but not before I have completed my "writing session"!

A word of caution at this point is in order about multitasking. It is commonly believed that multitasking is a skill to cherish. This may be true for some of our activities: ultimately it is nice to be able to listen to the radio while driving, or talk to a friend while walking in the park. However, when it comes to highly demanding tasks such as drafting some academic text, trying to do too many things at the same time will hinder your progress. My advice would be for you to focus on one specific task at a time, at most while listening to some classical music in the background.

Once you have managed to take action, to establish a routine, and to keep distractions under control, it is important that you keep a record of your progress so as to maintain your motivation. The following tool is designed to help you do precisely that. Simple as it may look, just give it a try, consistently over a few weeks and then observe your productivity soar.

1.5.3 Staying on track: Weekly check-ups

When I started my self-employment business around helping students and early career academics with their writing, I came across an online programme by Christine Kane [4] on how to set up a business. One of the many tools she shares in her programmes is what she calls a "Sunday Summit Form". The tool consists in a series of questions to help you reflect on your past week's achievements and plan for the next week's tasks. It can be used on any topic and it works in many different areas.

ACADEMIC WRITING WEEKLY CHECK-UP

1. **What have I accomplished this week?**
   Make sure you include everything you have accomplished no matter how big or small.

2. **Is there anything I wanted to accomplish but did not?**
   Be specific and include projects/items that have dropped out of the list of things you wanted to accomplish.

3. **What useful insight about my writing have I learned or experienced this week?**
   Pay attention to important clues about your writing routine. For example, note how long it takes to complete a task compared to how long you *thought* it would take. Find out which aspect of writing comes more easily and which one does not. Can you include something in your routine that makes your writing more enjoyable?

4. **What challenges am I experiencing?**
   If some specific challenge comes up over and over again, it may be because the task is either too big (if so, you need to break it down into smaller, more manageable parts) or not really important (if so, you should consider scrapping it altogether from your list).

5. **If I were to give myself advice, what would I tell myself about these challenges?**
   We are often far better at giving advice to others rather than to ourselves. Pretend a good friend comes up to you with the same challenges you are facing. What piece of advice would you give him or her?

6. **What are my top priorities for this coming week?**
   Planning the work ahead and being clear about your priorities is an effective way to keeping on track. Be realistic though with what you can achieve in one week. Only list three to four items and focus your attention on those.

7. **If I could get nothing else done this week but ONE THING, which one would I choose to do? Which one thing would make me happy and proud?**
   Well, this one is self-explanatory, really. Pick up that one thing and stick to it until completion. A sense of achievement is the best propeller forward.

This worksheet is based on the trainings of Christine Kane. Christine Kane is known as the Mentor to People Who are Changing the World. She is the president and founder of Uplevel You[TM], a multi- million-dollar company committed to the growth and empowerment of entrepreneurs and creatives around the globe through teaching not only high-level cutting-edge authentic marketing and business strategies, but also transformational techniques to shift mindsets and wealth. www.ChristineKane.com

One day, I decided to use the tool for my own writing. At that time I was struggling with making a start with a long-due research paper. All data had finally been analysed by my PhD student and we had agreed that I would prepare the first draft of the paper while he started writing up his thesis. Sadly, I never seemed to be in the right mood, until eventually I resolved to use this little tool I had learnt from Christine. Within three weeks, I managed to circulate a final draft to my colleagues for feedback and comments; a couple of months later the paper was accepted and finally published in a high-impact journal!

This success story is not isolated. I have since shared with students at my workshops a revised version of the tool and those who have regularly made use of it rave about its efficacy. An electronic copy of the tool provided here can be downloaded at: https://www.crcpress.com/9781498701471. Alternatively, you can decide to create your own version of it possibly re-wording some of the questions in a way that most resonates with you.

Unlike more traditional to-do lists, the weekly check-up has three key features that can really help you make progress with your writing.

First, it forces you to take some time to gather your thoughts and reflect on your performance so far. Specifically, it asks you to recognise and reward yourself for any big or small achievement or breakthrough. This is a much overlooked aspect in our routine, but since it is a crucial point, I have allowed myself a little digression in the focus box.

Second, the tool helps you identify any specific task that you are especially struggling with. This is no small insight as it indicates where your greatest blockages lie. Becoming aware of what they are is often the first and most important step before we can start tackling their root causes.

Third, it helps you in becoming strategic about your priorities. All too often, new tasks or commitments land unexpectedly on our desks. We then need to make adjustments to our plans in order to deal with the new situation that has arisen. Yet, if we have already identified our top priority for the week ahead, we can try and make sure that that priority gets done no matter what. It is reassuring and liberating to realise at the end of the week that, despite unexpected events and derailments, you have managed to complete the one thing that really mattered to you. In itself, this can be a major boost forward and a way to keep momentum in your plans.

Filling in the Weekly Check-Up form is not meant to become an onerous task and it should only take a few minutes once a week (I used to do mine on Sunday afternoons). Of course, it is up to you to decide how much or how little to write, but make sure you include relevant details wherever appropriate.

> ## On the importance of celebrating one's achievements
>
> We never seem to take time to recognise what we have *already* achieved. Without doing so, any new task ahead becomes like the famous carrot that lingers before our nose but that somehow we never seem to reach. Chasing a constantly moving target is frustrating and demoralising. Feeling like we never have enough time to do all the things we want to do becomes off-putting in the long term. So, rather than approaching any new challenge with a renewed zest, we just switch off thinking it is never enough.
>
> My mother once made me acutely aware of my awkward behaviour in this respect. I do not remember the exact occasion, but I guess I must have been upset and frustrated, probably for not having managed to do something I really cared about. As a result I must have made some melodramatic generalisation about myself, most likely along the lines of *"I am never going to be good enough at this"*.
>
> And that is when my mum completely changed my perspective. My problem – she said – was not whether I was or was not good enough, but rather that I never really acknowledged any of the things I had already achieved in my life. Somehow – she added – I acted as if the very fact I had achieved something meant that it was neither so important nor so difficult after all. This – she concluded – had been the case when I got a degree with the highest grades, when I moved on my own to a different country, when I learned to speak English, and then German, and then Spanish... the list went on.
>
> I suddenly realised that she was right, and decided that from then on I would start giving myself recognition, acknowledge and celebrate the things I had managed to accomplish even though plenty more may still lay ahead. You should do the same!

1.6   YOUR WRITING SKILLS: SELF-ASSESSMENT CHECKLIST

As you go through the steps and tasks presented in this book, you may want to assess your ongoing knowledge and practice in academic writing. For this reason I have devised a simple and quick checklist that you can use throughout your work to assess the progress you make. All you need to do is to decide to what extent you agree with each of the statements in the checklist, using a score of 1 (low), 2 (medium), or 3 (high). If you do not agree with a statement, simply assign zero to it. Your total score will give you an idea of where you are in the learning curve of good academic writing. Monitoring your progress while reading this book and putting into practice the strategies presented will help you find out how much you have learnt on the way. That in itself can be a good motivator to keep working at honing your skills.

Self-Assessment Checklist

1. I am well organised and know how to locate all my sources [ ]
2. I do not suffer from procrastination [ ]
3. I can easily extract key information from the papers I read [ ]
4. I am able to critically evaluate the work of others [ ]
5. I can easily identify the underlying structure in what I read [ ]
6. I am good at generating ideas and can write first drafts quickly [ ]
7. I spend enough time revising and editing my text [ ]
8. I regard feedback as an opportunity to improve my writing skills [ ]
9. I am familiar with most technical terms in my research area [ ]
10. I have a good command of the English language and grammar [ ]
11. I am familiar with bibliographical conventions in my research field [ ]
12. I carefully check my text for spelling and grammar mistakes [ ]
13. I can easily proofread my own text [ ]
14. I know how to produce good quality figures and tables [ ]
15. I believe being able to write well is an essential skill to master [ ]

Total [ ]

TOTAL SCORE AND PROFILES

0-15: Scientific Academic Writing is probably very new to you, either because you have had no experience or because you are still confused as to how to improve on previous unsuccessful attempts. This book will show you the specific steps to follow to make a start and move through the various stages of the writing process. If some of the material covered in this book feels a bit overwhelming at first, do not despair! Each step builds on the previous one and is designed in a way that will remove a sense of overwhelming and allow you to gain confidence as you go along. By following the steps, taking action, and completing the tasks assigned, you will soon start to enjoy your writing and make good progress.

16-30: This book is spot on for you! Whether you have already had some training or exposure to scientific academic writing, or whether you are naturally gifted at least in some of the skills required for good academic writing, this book will help you fill any remaining gaps. You will further develop your existing skills as well as a new degree of awareness and appreciation of good practice in scientific academic writing. With time and practice, you will be able to master your writing skills like an experienced academic.

31-45: Congratulations! You are obviously doing very well and there are probably only very few areas where you need to improve. You may already be familiar with some of the material presented here. This book will help you consolidate the things you do well already, while at the same time providing tools, tips and strategies for a higher level of professionalism. It is my hope that it will finally transform you into an excellent academic writer in your discipline.

Chapter 1: Before You Begin
In a nutshell...

- Scientific academic writing is a craft and it can be learned.
- Approach writing as you would any other commitment: set aside time to write and protect it from other people's demands on your time.
- Ban distractions (yes, these include Facebook, Twitter, email pop-ups and other notifications).
- Use time framing to decide how much time you wish to devote to a specific task. Start with short amounts of time (15-20 minutes) at first and increase them gradually.
- Use regularly the Weekly Check-Up form to assess how well you are doing and keep momentum.
- If it helps, find a writing buddy or join a writing retreat.

EXERCISES

1.1 **Academic Writing Check-Up.** Download a copy of your form at: https://academiclife.coachesconsole.com/downloads.html or https://www.crcpress.com/9781498701471
Spend a few minutes once a week to fill in the form and review your progress after a few weeks.

1.2 **Self-Assessment.** If you have not done so already, complete the self-assessment checklist and identify the areas of greater concern to you.

FURTHER READING

http://pomodorotechnique.com

http://www.wikihow.com/Stop-Procrastinating

http://www.marcandangel.com/2010/11/22/7-common-causes-and-proven-cures-for-procrastination/

II

THE WRITING PROCESS

CHAPTER 2

Pre-Writing Step

CONTENTS

2.1	Gathering and organising information		20
2.2	The three objectives of reading		21
	2.2.1	Capturing content: Taking notes while reading	21
		2.2.1.1 The paper annotation tool	22
	2.2.2	Capturing structure: Developing a template	24
	2.2.3	Capturing style: Learning from the masters	25
2.3	The literature review		26
	2.3.1	What it is and why it matters	26
	2.3.2	The literature review as a process	27
	2.3.3	Critical voice	27
	2.3.4	The literature review as a product	31
	2.3.5	How to write a literature review	32
	2.3.6	Literature review matrix	33

INFORMATION has never been so readily available as it is today. Gone are the times when scholars used to spend hours upon hours in a library searching for a paper relevant to their research. Today, most papers are available online, either through institutional subscriptions or via open access repositories for direct download.

Papers are not the only sources of information for a researcher, though. Others include books, theses, dissertations, conference proceedings, lecture notes, laboratory activity reports, collaboration reports, grant proposals, scientific magazines, outreach publications, and more. In fact, the risk is often to be in a status of constant information overload.

Becoming familiar with the amount of information available on any given topic can be a daunting prospect for many PhD students and early career researchers, especially when it comes to summarising previous studies into a literature review. Thus, it is important to be systematic in the way the information is sourced, stored and retrieved.

Even more important, however, is learning to *read* these sources and to extract valuable insight not just on content, but also on structure and style so that one is better prepared to write well.

2.1 GATHERING AND ORGANISING INFORMATION

Gathering and organising information should be one of the primary tasks of a conscientious student from the very early stages of their PhD project. It is also a key objective for any academic approaching and progressing through a new project.

Nowadays, many databases exist that can be easily accessed online for quick retrieval of papers in virtually any discipline. If you are unsure on how to use them, you may wish to contact local librarians who are normally very willing to help.

Once you have located sources of information potentially relevant for your project, you may want to store them in a way that makes it easy for you to retrieve them in the future. This can be done with any of the reference management software available, such as Endnote, Mendeley, Papers, Zotero[1]. As you begin your PhD you may wish to spend some time familiarising yourself with the package that best suits your needs. However, I would advise you against spending too much time trying to work out their full potentialities or constantly looking for the latest new kit. What matters here is that you devise a personalised approach that works well for you.

When I was a PhD student, I used to download electronic papers and store them on my computer by the name of the first author and the year of publication (e.g., smith-1998). You can be more elaborate than that and add the journal name or use tag words that indicate the type of paper (whether it is theoretical, experimental or computational), or its key topic, or any relevant information that quickly reminds you about the content of the paper. Again, the aim here is to easily retrieve any information of interest even months or years after you first stored it.

As you build up your collection of papers (and other sources) it is also advisable to keep a record of every item in your 'library' by creating a file with all relevant bibliographic details. If you use LaTeX [5, 6] as your main editing tool (which I would strongly recommend for excellent production quality), you can create a BibTeX [7] file as a reference repository for any of your writing project, whether your PhD thesis or your research papers. Several books and online guides are available on the use of LaTeX and its associated packages, if you are not familiar with them.

Clearly, sourcing and storing papers and relevant articles is a process that continues throughout your research and you should get into the habit of staying abreast of the vast amount of literature produced as you progress with your project. However, storing your sources of information is only half the job. Now you need to read them and learn how to make the most of them!

---

[1]Discussing them here goes beyond the scope of this book, but you can easily compare their features through a quick Google search (see for example: https://en.wikipedia.org/wiki/Comparison_of_reference_management_software).

2.2 THE THREE OBJECTIVES OF READING

When students ask me how to become better academic writers, my answer is always the same: by becoming a better reader first! Indeed, any writer is first and foremost a reader and there is so much that one can learn about writing by focussing not just on *content* but also, crucially, on *structure*, and *style*.

These are the three key objectives of reading: mastering each is essential to becoming a better reader and, in turn, a better writer.

2.2.1 Capturing content: Taking notes while reading

The first thing we focus on when reading a paper, a thesis, or a report, is typically *content*: is the information relevant to us? Do we get the answers we were looking for? Do we agree with the opinions expressed? Yet, we do not always capture content in a systematic way and we easily forget about what we have read, even a few months later.

One of the most effective ways of capturing content from the material we read is by taking notes. This is best done on paper rather than electronically. Annotating papers by hand (either at the margins of the paper itself, or on separate sheets) helps to crystallise information in our brains better than if we stored our notes on a computer. Of course, I would not argue you should print every paper you come across. But for those ones that are really important to your project, you should definitely consider printing them out and taking hand-written notes on them.

Deciding which paper is important can be done through some quick skim reading. The following focus box provides some useful tips on how to take notes while reading. Once you have familiarised yourself with the content and understood most of it, you can proceed to filling in the Paper Annotation Tool, as explained in the next section.

> ## Tips on Taking Notes when Reading
>
> When reading a paper or a thesis chapter, you are more likely to remember its content if you take notes as you read. Here are some useful tips for effective note-taking:
>
> - Print out a copy of the source you are reading (a thesis chapter, a research paper, a collaboration report).
>
> - Underline key concepts or results as you read.
>
> - Highlight any factual information you are likely to need again (e.g., for your literature review). This can be a final numerical result; an outstanding issue; a key reference to some other source.
>
> - Take explanatory notes as you read. Try to be concise and to the point. If possible, write at the margins of the paper. Alternatively, use a separate sheet of paper and make sure you keep it physically attached to the source it refers to.
>
> - Whenever possible, take notes by hand NOT on the computer. Writing longhand will help you memorise better what you have read.

2.2.1.1 The paper annotation tool

Many PhD students (and indeed most of us) tend to forget the content of what they read, especially after having read several dozens of papers a few years into their doctorate.

A useful way to overcome this problem and to standardise the notes you take is to use what I call a *Paper Annotation Tool*, or PAT in short. It consists in a single double-sided A4 sheet containing a set of pre-defined and general questions with some space for answers. A copy of the PAT I normally use is reproduced here (you can download an electronic copy at https://www.crcpress.com/9781498701471 or or create your own). Just make sure you fill in the PAT only after having fully read and understood the paper.

The questions are largely self-explanatory and provide a versatile tool designed to help you extract and record key content and bibliographical details of any paper you read. If you work in a laboratory-based discipline, many of the papers you read will be of experimental nature and you can use the template provided as shown. If your discipline is more theoretical or computational you can re-phrase some of the questions on the sheet to better suit your needs.

Paper Annotation Tool

TITLE:
JOURNAL:
AUTHOR(S):
VOLUME:
YEAR:
PAGE(S):

What is the paper about?

What is the aim of the study?

Why is it important?

What is the approach/method used to acquire the data?

What is the approach/method used to analyse the data?

What are the key findings?

Is there any limitation?

What are the main conclusions and implications in the wider context?

Any other comments?

(feel free to modify any of the questions listed to fit your research area)

Whatever your discipline, the main purpose of using a PAT is to focus your mind on specific key aspects of the articles you read. Answering the questions in the PAT may take you longer than expected at first. However, the process is not meant to be tedious or overly time-consuming. In fact, with some practice you should be able to fill in the PAT sheet in just a few minutes.

As you keep using the PAT, you will gradually notice that the way in which you *approach reading* will change, as you will mentally remember the questions in the PAT and be in a better position to focus on what is important about the paper while you read it. In other words, you will start paying attention to things you might have previously overlooked, such as any limitations of the study or any flaw in the statistical significance of the results presented. With time, filling in the PAT will turn you into a more critical reader and help you develop your own critical voice (Section 2.3.3).

In Section 2.3.6, we shall see how a systematic use of the Paper Annotation Tool can also prove extremely useful in planning and creating a matrix for your literature review.

2.2.2 Capturing structure: Developing a template

PhD theses and research papers in scientific disciplines follow a well-defined structure with key and clearly signposted chapters or sections. These are normally referred to as *Introduction, Methods, Data Analysis and Results, Discussion and Conclusions*, even though their titles, or indeed their order, may vary across disciplines or type of article (for example those appearing as a *Letter* or a *Rapid Communication*, rather than a *Regular Article*).

What is perhaps less obvious to many inexperienced writers is that even individual chapters, sections, and paragraphs obey to a standard (almost predefined) structure, where information is presented to the reader in a very specific sequence.

As the structure provides the skeleton without which the whole document would fall apart, understanding how to structure individual parts of a research paper or thesis is vital to writing well and learning to capture this structure is another crucial aspect of reading.

The best approach I found about uncovering the structure of individual sections is the one suggested by Hilary Glasman-Deal in her excellent book *Science Research Writing for Non-Native Speakers of English* [8]. The approach consists in focussing not on the meaning (i.e., content) of each sentence, but rather on the *function* that each sentence accomplishes within a paragraph. In other words, understanding what the writer is trying to achieve with each sentence, and why these are presented in a given order, can allow us not only to recognise a distinct pattern in individual sections (*Introduction, Methodology, Discussion, ...*), but also – more importantly – can allow us to create section-specific templates to follow when we start writing.

The process outlined here will be fully presented in Chapter 9, together with a specific example on how to carry out a thorough analysis of the structure of an *Introduction*. Templates for the most common sections of a research paper or thesis are also presented in Chapter 9. Ideally, you should spend an adequate amount of time to familiarise yourself with the intended structure of each section of a typical paper in your own discipline before writing any content.

If you are reading this book while trying to write a paper or thesis, I would recommend that you begin with an individual paper section or thesis chapter first, using its corresponding template provided in Chapter 9. Focussing on the same section/chapter throughout, apply the step-by-step framework presented in this book and only move to a new section/chapter once you have completed all the steps. This will give you a good overview of the type of work (and time!)

required to produce a well-written final version. Hopefully, you should notice that writing new sections/chapters will prove easier and easier as you apply the approach presented here.

2.2.3 Capturing style: Learning from the masters

When I was an Alexander von Humboldt Fellow in Germany, I used to marvel at the manuscript drafts of my host, Professor Claus Rolfs, a world-leading figure in Nuclear Astrophysics and author of the book *Cauldrons in the Cosmos*, for many decades regarded as the *bible* in the field. I remember how his writing seemed to come out effortlessly and how everything was always concisely yet clearly expressed. I wish I had his talents! In an effort to improve my own writing style, I consciously started to pay attention to the *how* and not the *what* of his drafts. And to my great surprise, I started realising there was so much to learn by simply emulating what he had written.

Today, when I run my writing workshops I tell my students that an excellent way of improving on their own writing is by paying attention to the way of the masters. The easiest way of doing so is through... *copying*! Of course, I do not mean plagiarising material written by others. Rather, I mean using long-hand copying merely as an exercise to focus more closely on details that you may otherwise miss if simply reading or copying using a keyboard. When writing longhand, you are forced to slow down unless, of course, you are faster at writing by hand than you are at typing. Slowing down helps you notice the nuances of the text. And that is precisely where style hides in writing.

So, here is a simple task for you to start learning about style: find a paper (or indeed any text) that you regard as being very well written; focus on a specific excerpt (a paragraph, a sentence, a figure caption); and start copying it by hand. Incidentally, this exercise can be an excellent antidote to the writer's block (Section 3.2): by spending a couple of minutes copying someone else's text as a way to 'get going' you may soon realise that you have found that magical inspiration that sets you off to your own writing.

As you copy your targeted text, pay attention to the way sentences relate to one another: connectors such as *however*, *although*, and *therefore* are very powerful ways of alerting the reader to a change of direction (or indeed the reinforcement of one). However (!), more subtle links can be created by overlap (repeating something said in previous sentences); pronouns or relative clauses; or even the wise use of punctuation. Pay attention to the length of each sentence and how this can be used to provide emphasis: often a very short sentence after a rather long one acts like a punch in the stomach to attract the reader's attention! Notice how verb tenses are used. For example, reporting on research already carried out is typically done using past tenses. At times, however, the present tense is used instead. Ask yourself: why is that? Is the author trying to imply something without saying so explicitly? Try and pick up the subtleties of the language by reading between the lines.

The 'copying-by-hand' exercise works just as well for a poorly written excerpt. In fact, you may exploit this technique to discover precisely *where* the problems lie in a text that you regard unclear or inelegant. You may realise, for example, that sentences are too long or too convoluted; that the subject is fifteen words away from its verb; that some pronouns are left hanging in the air without being clearly related to any other element of the sentence. Once you become aware of such pitfalls, just ensure that you do not make the same mistakes in your writing.

We often tend to forget that any piece of good text we read may have taken its author a considerable amount of time and effort. For example, a well-written paper in a peer-refereed journal may have undergone several iterations before appearing in its final form. Emulating to the way in which other writers render specific facets of scientific academic writing (accuracy, impartiality, objectivity, rationality) is especially helpful to develop your own style and hone your skill set. After all, good academic writing is an art and like any other form of art it takes time to master.

2.3   THE LITERATURE REVIEW

For most PhD students, writing a Literature Review is one of the earliest tasks they face when it comes to writing. Often, this is also one of the most daunting prospects for many scholars.

2.3.1   What it is and why it matters

The phrase 'literature review' is used to indicate both a *process* and its *outcome*. The process of carrying out a literature review consists in searching all papers relevant to your study. The search starts at a very early stage and shapes up the research question(s) that you plan to address in your project. It also helps identifying the key authors and journals relevant to your research.

Searching the literature and keeping up to date with the latest developments in our fields typically continues throughout the lifetime of a project and eventually culminates in the production of a literature review, a key component of any thesis or research paper. Intended as the outcome, the literature review thus provides a comprehensive overview of past and recent knowledge on a given topic and illustrates the context within which your own research is located.

Both the process and the product of a literature review are crucially important aspects of any research and it is imperative to do both well. Yet, it is not unusual for many to start reading papers upon papers only to realise that they have no clue as to *how* to condense in a few pages all the information they have unveiled.

The following sections offer a few key pointers to help you with both the process of performing a literature review and the process of *writing* one. For more in-depth overviews of all aspects related to literature reviews, you may wish to consult *The Literature Review. A Step-by-Step Guide for Students* by Ridley [9], or *The Literature Review: Six Steps to Success* by Machi and McEvoy [10].

2.3.2 The literature review as a process

Searching the existing literature is greatly facilitated these days by the availability of powerful tools such as Google Scholar and extensive databases such as Web of Knowledge, Scopus, inSPIRE, MEDLINE, and EThOS. These should quickly help you locate *review articles* in your discipline. Review articles offer an excellent starting point for a literature search as they provide an exhaustive reference list and guide you through the latest developments, ideas or trends in your field. Other sources include *grey literature* (theses, laboratory reports, conference proceedings and posters) as well as field-specific webpages and professional organisations.

Once you have located a number of sources that may potentially be relevant to your research, you need to decide *what* to read and *why*. In particular, it is important that you establish the scope of your search early on so as to avoid becoming overwhelmed by the huge amount of information available.

Not all papers will be equally important to you. Start by simply scanning the content to find sections or keywords that provide the information you are seeking. Then proceed with skimming, namely reading through parts of text that give you an overview on content. And finally, move to in-depth reading from beginning to end. Just make sure this latter activity is only devoted to the papers that really matter to your project. These should ideally be printed out and stored with their respective Paper Annotation Tool sheets (Section 2.2.1.1).

2.3.3 Critical voice

A key aspect of carrying out a literature review consists in critically assessing the work of other scholars. For this purpose, you need develop your own critical voice which implies: looking for evidence that supports a claim; checking that the authors' conclusions follow from sound, logical arguments; assessing what implicit or explicit assumptions have been made, if any; verifying whether the authors' claims match those of other scholars or indeed your own evidence or knowledge.

In my experience, one of the best ways to improve personal critical skills consists in joining a Journal Club, namely an informal gathering of scholars who meet to discuss and critique a selected paper. If no journal club is available in your department, consider creating your own. All you need is just a couple of colleagues (other PhD students, young researchers, or more senior academics) with research interests similar to yours and the agreement to meet regularly (ideally once a week) for an hour or so.

The focus boxes that follow offer some tips on how to start a journal club and what items should typically be discussed. Briefly, people within the club take charge in turn to select a target paper and circulate it to the other members some time before the meeting. During the meeting, one person presents the paper and leads the discussion, soliciting opinions from the other members. Questions that may be worth addressing include the validity of the approach(es) used, the novelty of the contribution, any potential limitation of the study, and so on. You may decide to circulate a set of common questions before the meeting or leave it to individual participants to bring their own.

In addition to improving your ability to evaluate the work of others, being part of a journal club can also foster your confidence, improve your argumentation skills, and promote a greater sense of belonging to your academic community. Last but not least, it will also teach you how to anticipate the expectations and potential objections that your readers may have: an outstanding vantage point when it comes to your own writing!

Tips on Starting a Journal Club

Even if journal clubs are most effective with at least 4-5 members, one other person is all you need to make a start. This can be a fellow PhD student or a post-doctoral fellow. They do not have to be from within your research group, but ideally you want someone from the same discipline as yours.

Once you have found a willing partner, here is what to do:

Before and during the meeting:

1. Agree on a date and time for the meeting to take place. Ideally, once a week for one hour at most.
2. Identify a facilitator for the meeting. This is a role that can rotate amongst the members of the Journal Club.
3. Select a paper for discussion. This is normally done by the facilitator, but anyone can make suggestions.
4. The facilitator circulates the paper at least two days in advance of the meeting, possibly with a list of questions that will form the basis for the discussion.
5. Each member commits to reading the paper before the meeting.
6. At the meeting, the facilitator provides a brief oral presentation of the paper under discussion. There is no need to prepare slides or anything; a piece of chalk and a blackboard should be enough to write down key points if needed.
7. Everyone contributes to discussing and critiquing the paper (see the next focus box for possible questions to address).
8. Before the meeting ends, participants agree to a new date/time and identify the facilitator for the following meeting.
9. Meetings should take place regularly and start and finish at the agreed times.

ITEMS FOR DISCUSSION AT A JOURNAL CLUB

When discussing a paper, you can choose to focus on all or some of the following aspects (feel free to add your own):

Description of the Study

1. What was the purpose of the research?
2. Why is the research important in the wider context?
3. What was the nature of the study (experimental, theoretical, computational)?
4. Were its key objectives clearly stated?

Literature Evaluation

1. Was the literature review well presented and sufficiently up to date?
2. Was any major recent study left out?
3. Is the paper clear and well written?

Approach and Analysis

1. What was the method used in the study? Can you clearly identify it?
2. How were data obtained and analysed?
3. Is/was there any fault in the approach used?
4. Is the statistical analysis of the data appropriate and sound?

Results and Conclusions

1. What were the key findings of the study?
2. Were results well presented and properly discussed?
3. Did the author(s) provide an interpretation of the results?
4. Did the author(s) discuss any potential limitations of the study?
5. Could the study be replicated?
6. Was the study successful in solving the research gap(s) identified?
7. What additional questions does the study raise?

2.3.4 The literature review as a product

The outcome of an extensive literature search culminates in a compendium of the latest advances in your topic known as the *Literature Review*. Whether part of a PhD thesis, a research proposal, or a research paper, the literature review fulfils several key functions, namely:

- it brings the reader up to date with current knowledge on a topic;
- it locates a *gap* in research;
- it provides a *justification* for undertaking the study (this is especially important in applications for funds, given that the mere existence of a research gap does not necessarily imply it is important to address it);
- it demonstrates your knowledge on the current status of a field;
- it provides evidence (through citations) to support your claims, thus affording you a greater credibility.

Your research does not exist in isolation. Rather, it represents a tile in a much bigger jigsaw puzzle put together by what other scholars have done before you. Thus, the literature review provides the context for your research project and summarises previous studies as the foundations on which your own research builds. Depending on your specific field or topic, your literature review will take up one of four possible forms: traditional, systematic, meta-analysis, or meta-synthesis.

A *traditional* (or *narrative*) review is probably the most common type. It provides a summary of a selected body of literature as a comprehensive background on a current research topic. A *systematic* review, by contrast, provides a rigorous and well-defined approach to review *all* literature in a given subject area. A *meta-analysis* review represents a statistical analysis on a large body of quantitative finding; whereas a *meta-synthesis* consists in a non-statistical analysis to integrate, evaluate, and interpret findings from qualitative studies. Typically, reviews in scientific subjects will be of the first type, with the others being more often used in social sciences and policy making. It is important that you identify the type of review you are going to write, as this will also frame the extent of your searching. If you are unsure, discuss things with your supervisor(s) or find out what is typical in your own discipline.

Unlike other sections or chapters of your thesis (or research paper), the literature review does not have a standardised format. In fact, even its location within the main body of your work may vary depending on the nature of your study and/or your discipline. Often the Literature Review may appear as a separate chapter in your thesis (sometimes referred to as *Scientific Background*). In some cases it is outlined in general terms in the *Introduction* chapter and revised in greater detail at the beginning of subsequent chapters; or it may appear interspersed throughout the whole thesis. There is no right or wrong approach, but the right location may depend on the nature and purpose of your study. So, it is important that you find the format that best fulfils your needs.

Regardless of where it appears in your document, a well-written literature review will contribute to earn you authority, validate your choice of approach, and argue the case for the research statement or research gap that your study intends to address. While there is no pre-defined format for a literature review, it helps to arrange the material in a way that makes it easy for your readers to follow.

In science, recent and latest advances are typically mentioned first. However, depending on your project, you may need to trace the development or progression to current thinking and a more chronological order may be needed. Other common approaches in structuring your review include moving from the general to the specific, and comparing and contrasting. Other possible structures are discussed in Section 4.3.

2.3.5 How to write a literature review

So far we have discussed the many facets of a literature review and its importance. But how do you actually write one? Given the extent of the undertaking, the best way to approach the task is to organise your work around the following successive steps:

1. Identify the research question your study aims to address (and hopefully answer). If different questions are addressed, elaborate a research statement that encompasses them all under a common overarching theme that provides a focus for your writing.

2. If applicable, break down individual components of your research question. For example, consider all the necessary aspects required to demonstrate a thesis (research statement). Hence, organise your material (and relevant papers) around them.

3. Consider the order in which you need to present the information to your readers based on the knowledge they already possess (Section 3.1).

4. Map out the main point of each paragraph and support each with relevant references. Keep revising the order of ideas until everything flows in a logical way. Remember that each step builds upon previous knowledge.

5. Keep a clear writing direction: whenever you struggle to write, take that as an indication that you may have to do some more work to create a detailed plan of action.

In the next chapter, we will explore a couple of techniques to help you organiser your thoughts in preparation for your writing.

For extended reviews, decide whether you should organise your material by topic or theme and also make sure you include an introduction and a conclusion section (or paragraph) to summarise the key points you want your reader to remember. Where appropriate, use section headings to indicate a shift in topic and to provide a visual break to your text (see more on the structure of paragraphs in Chapter 4).

Citing other people's work. When citing other people's work, use paraphrasing rather than direct quotations and always provide a reference on any results, graphs, theory, or ideas sourced from someone else. In general, any claim or statement must be backed up by sufficient evidence, except if it forms 'common knowledge' in the field. So for example, you may write: **nuclear reaction cross sections drop by orders of magnitude at sub-Coulomb energies** without any specific reference because this is common knowledge in nuclear physics[2], but you would have to add a reference to a statement such as: **the cross section was 0.3 μbarn** as it refers to a specific measurement.

On the other hand, if you are writing a report as part of your academic assessment, you may want to cite references also for aspects of common knowledge of your readership as a way of showing to your assessors your familiarity with the topic and your ability to locate textbook information and other sources.

The format you use to reference other people's work will largely depend on your discipline. The two most commonly used formatting styles in science are discussed in Chapter 6.

2.3.6 Literature review matrix

As we have seen, one of the key purposes of a literature review is to present a critical overview on the current state of knowledge on a given topic or research area. Sometimes, this can be a difficult task especially when previous work has been extensive. In such cases, preparing a *literature review matrix*[3] can prove invaluable. The matrix is merely a table that lists a series of specific research questions or aspects of interest and their corresponding "results" as obtained in previous studies.

An example of a matrix I prepared for the literature review section of a research paper I was drafting is shown in Figure 2.1. The paper was about the study of a nuclear reaction important in various astrophysical sites. Even though the reaction had already been studied by different groups and with different approaches (including some theoretical calculations), the overall agreement remained poor and the reasons for discrepancies unclear.

In order to facilitate the task of critically comparing and contrasting different results, I prepared the matrix shown in Figure 2.1. Once I had all previous

---

[2] Unless of course, your readership is from a broader background than nuclear physics.
[3] I first came across this term on the excellent blog *The Thesis Whisperer* (http://thesiswhisperer.com), but many other similar phrases exist.

	Direct Capture contribution	Method	Energy range [keV]	Resonance strength (183 keV)
Rolfs (1973)	constant S-factor	prompt gamma-ray detection	280-425	not measured
Fox *et al.* (2004)	not measured	prompt gamma-ray detection	resonance	$(1.2\pm0.2)\times10^{-6}$ eV
Fox *et al.* (2005)	questions Rolfs' data: may have been affected by contributions from two broad resonance tails. S_{DC} ~2.5x less than in Rolfs	realistic Woods Saxon potential	resonance	$(1.2\pm0.2)\times10^{-6}$ eV
Chafa *et al.* (2005, 2006)	not measured	activation	resonance	$(2.2\pm0.4)\times10^{-6}$ eV
Newton *et al.* (2010)	S_{DC} in agreement with Fox's predictions. S_{DC} ~2x less than in Rolfs	prompt gamma-ray detection	257-470	not measured
Hager *et al.* (2012)	S-factor in good agreement with Newton, but higher than Fox	DRAGON recoil separator	260-470	not measured

Figure 2.1 Example of a literature matrix showing key research aspects (columns) and their corresponding outcomes (rows) as reported in various papers. A literature matrix can simplify the task of critically comparing and contrasting existing literature.

values and outcomes at a glance before my eyes, writing a literature review paragraph for the paper became very easy. It also highlighted how the different systematic uncertainties involved in the use of different experimental approaches (in this example, **activation measurement** and **prompt gamma-ray detection**) might have been the reason for the discrepancy (about a factor of 2) between the **resonance strength** values reported by Chafa *et al.* and by Fox *et al.*, respectively. An example of a possible literature section based on this matrix is shown in the following box.

The beauty of a literature review matrix is that is can be done at different levels of sophistication and for any topic. If the extent of your literature review is too broad, consider using different matrices, possibly arranged by theme. Of course, if you have filled in a Paper Annotation Tool (see Section 2.2.1.1) for each of the papers that you wish to include in your review, you should be able to quickly summarise all key information in your matrix and critically assess the current state of affairs in your review.

EXAMPLE OF A LITERATURE REVIEW SECTION

Early investigations of the ^{17}O(p,γ)^{18}F reaction cross section had reported a constant direct capture (DC) contribution to the S-factor (S_{DC}) in agreement with the four lowest data points measured by Rolfs (1973) at energies $E = 280 - 425$ keV by prompt γ-ray detection.

Fox et al. (2005) later questioned whether these data points were dominated by the DC process or rather affected by the presence of the two broad resonance tails. They thus calculated the S_{DC} using measured spectroscopic factors and a realistic Woods-Saxon potential, obtaining an S_{DC} value up to a factor of $\sim$2.5 lower than that reported in (Rolfs, 1973). This result was later confirmed by Newton et al. (2010) through prompt γ-ray measurements in the energy range $E \simeq 257 - 470$ keV.

More recently, the total (i.e., DC plus broad-resonance contributions) S factor was determined at $E = 260 - 470$ keV using the DRAGON recoil separator at TRIUMF (Hager et al., 2012) and found to be in fairly good agreement with values by Newton et al. (2010), albeit consistently higher than the total S factor reported in Fox et. al. (2005).

As for the $E = 183$ keV resonance strength, only two values exist to date: $\omega\gamma = (1.2 \pm 0.2) \times 10^{-6}$ eV, as determined by a prompt γ-ray measurement (Fox, 2004, 2005), and $\omega\gamma = (2.2 \pm 0.4) \times 10^{-6}$ eV, as determined by the activation technique (Chafa et al., 2005, 2006). They disagree at the 95% confidence level. The origin of this discrepancy is not understood at present, but may be due in part to unobserved gamma transitions in Fox et al. (2004, 2005) and/or an inappropriate subtraction of the DC component in either Fox et al. (2004, 2005) or Chafa et al. (2005, 2006).

Unfortunately, the lack of experimental data at low energies and the largely unconstrained S_{DC}-factor have so far precluded the accurate determination of the thermonuclear rate for the ^{17}O(p,γ)^{18}F reaction. A renewed effort for the experimental investigation of this important reaction is thus required.

Chapter 2: The Pre-Writing Step
In a nutshell...

- Become a better reader by paying attention not just to *content*, but also to *structure* and *style*.

- Get in the habit of taking hand-written notes on the papers you read, especially the ones critically relevant for your project. For the latter, fill in a Paper Annotation Tool sheet. This will be useful for later retrieval of information and for creating a literature review matrix.

- Learn to develop your critical voice by joining a journal club. If none is available at your department, consider establishing one yourself. Guidance on how to do so is given in this chapter.

- Familiarise yourself with the most common type of literature review in your research field and discuss with your supervisor the extent and purpose of your search.

- Plan your literature review carefully by arranging key elements by theme if necessary. Use a literature matrix to gain a comprehensive overview of the current status of knowledge in your project area.

EXERCISES

2.1 **Take notes** [10 minutes]. Using the Paper Annotation Tool, take notes for every important paper that you read. Keep the PAT attached to the paper and store in a suitable folder for later retrieval.

2.2 **Unveil the structure** [30 minutes]. Following the outline presented in Section 2.2.2, start paying attention to the underlying structure of different sections in the papers you read. In Chapter 9 we will explore more thoroughly how to develop section-specific templates.

2.3 **"Copy" from the masters** [5 minutes]. Take an excerpt from a paper or thesis in your discipline and start copying it by hand while paying attention to the *way* in which things are written. Refer to Section 2.2.3 for what you should pay attention to while copying.

2.4 **Prepare a literature matrix** [20-40 minutes]. Before you begin writing a literature review, consider preparing a matrix like the one in Figure 2.1. Keep adding to it as you find new papers and other sources. For extended reviews, consider preparing a matrix for each theme or sub-topic if necessary.

2.5 **Putting things into practice**. If you decide to start writing while reading this book, focus on one section/chapter at a time and cover all the steps involved in the writing process (from Pre-Writing to Proof-reading) before moving to a new section/chapter.

FURTHER READING

Wallace, M. and Wray, A. (2011). *Critical Reading and Writing for Postgraduates*, SAGE Study Skills - SAGE Publications Inc. (second edition).

Researching for your Literature Review – Monash University
http://www.lib.monash.edu
http://guides.lib.monash.edu/researching-for-your-literature-review/home

Library & Learning Commons – Bow Valley College
http://bowvalleycollege.libguides.com/literature-reviews

On all matters around doing a PhD
The Thesis Whisperer
http://thesiswhisperer.com

And for some fun
PhD Comics
http://phdcomics.com

CHAPTER 3

The Drafting Step

CONTENTS

3.1	Who is your audience?	39
3.2	Talking about your research	41
3.3	Getting the structure right	42
3.4	Mind maps	43
3.5	Core dump	47

BEGINNING to write is hardly an easy task for most of us. So, in this chapter I offer some useful techniques designed to make it easier for you to make a start. Not all techniques and approaches will appeal to everyone, but I advise you to try each of them at least a couple of times in order to see what works best for you. With some practice, these techniques will soon become second nature and you will automatically resort to them every time you start writing from scratch, whether you are working on a research paper, a thesis, or other academic reports. Ideally, you should apply these techniques in the order in which they are presented here, as each naturally feeds into the next. As you progress, you will move towards an increasing level of complexity with more writing gradually required.

3.1 WHO IS YOUR AUDIENCE?

The first and most important question you should ask yourself before you begin to write is: "Who is my audience?" If you cannot answer this question clearly and precisely, do not even attempt to start writing! Your readership is the single most important consideration to keep in mind before you begin.

Some people advise their students to "write for themselves", at least in their first draft. Indeed, some practitioners even admit to using this strategy. But I would disagree with this approach. As the person who carried out most of the work you intend to write about, you are far too familiar with the purpose of your study, the importance of your results, and all the steps you have taken to get from the original idea to the actual realisation of your project. Your reader does not share any of that! So the danger of "writing for yourself" is that you may end up writing something that only you are able to understand. As we have seen, this is not what academic writing is all about.

In my view, a far better approach consists in thinking carefully about your readership; ideally, you should spend some time answering all of the following questions:

- Who are your readers?
- What is their background?
- How familiar are they with your topic?
- What knowledge do they have about the current status of research in your field?
- What essential piece of information do they need at each point in your paper/thesis so that they can understand what follows?

Having answered these questions you will be in a better position to decide on the appropriate amount of details and information you need to provide and to pitch your language at the right level. Clearly, your audience will also depend on the type of document you are writing. So, let us consider in turn the case of a PhD thesis, a research paper, and a research grant proposal.

Writing a PhD thesis. If you are writing a PhD thesis, you are writing for a variety of different people with different backgrounds. These include your examiners, other scientists in your field, other students. Your examiners are likely to be experts in your chosen discipline, but may not necessarily have a deep knowledge or understanding of the *specific subject* of your project.

Other scholars may be interested in undertaking a project similar or related to yours. In this case, they are more likely to have a good grasp of the problem at hand, but may fail to have a broader knowledge of the wider context (for example, if they are themselves PhD students or early career researchers).

Finally, other students in your department (or indeed elsewhere) may or may not have any knowledge in your discipline, let alone in your specific project. Yet, they might be interested in seeing how another student (you!) has approached structuring his/her thesis, presenting their data analysis, arguing their conclusions.

Each of these readers will carry different background knowledge and expectations and somehow you should be able to cater for all of them at the same time. In the case of a PhD thesis, this can be achieved, for example, by:

- including specific chapters that summarise key aspects of the theory behind your work;
- providing appropriate definitions of any technical term used;
- adding appendices with further details on specific items.

In general, whenever addressing a mixed audience that comprises both experts and non-experts, always cater for the least knowledgeable of your readers.

Writing a research paper. If you are writing a research paper, then different considerations apply depending on the type of journal you wish to target. When submitting a manuscript to a *specialised journal*, you may safely assume that most of your readers will be as familiar as you are with the background of your research field and they will most likely be able to understand much of your jargon or technical language.

Different is the situation if you want to publish on a *high-impact journal* that typically targets for a wider audience. In such cases, you will have to make sure that you clearly explain any topic-specific terms and that you provide relevant background information to help your reader understand what follows.

You will also have to use an accessible language to convey the importance of your study in a way that non-experts can understand and to explain why it deserves the broader dissemination typically afforded by high-impact journals. Finally, you will need to achieve all this within strict page limits, not an easy task to master!

Writing a research grant proposal. If you are writing a research proposal either to get funds or to apply for a fellowship, your readership will likely be very heterogeneous. It is not uncommon for selection panels to comprise individuals from diverse scientific fields, from biology to astronomy, chemistry, particle physics and the like.

Clearly, it is not realistic to expect that each (or any!) of them will be familiar with your research area, let alone your specific project. Your task is then to use a language accessible to non-experts, albeit supported by a sound scientific background. In your proposal you should be able to share the importance of your research, your enthusiasm for it, and the reason(s) why it should be funded or, in the case of a fellowship application, why you are the best person for the post.

You can find very useful advice on how to write successful grant applications, in *The Research Funding Toolkit* by Aldridge and Derrington [11] and in *The Grant Writer's Handbook. How to write a Research Proposal and Succeed* by Crawley and O'Sullivan [12].

Now that you have identified your audience, you are a step closer to know how to pitch your writing. Yet, the act of writing may still feel a bit daunting...

3.2 TALKING ABOUT YOUR RESEARCH

I bet you have never heard about the *talker's block*. The reason, as Seth Godin argues in one of his popular blog posts [13], is simple: we are in the *habit* of talking. In fact, *we get better at talking precisely because we talk* [13]. It is as trivial as that.

With writing, however, things are different. When facing a white sheet of paper (or screen!) most of us feel a sense of resistance almost as if whatever we put down had to be perfect the first time round. We focus on getting it *right*, rather than getting it *done*. And this can become an insurmountable barrier for many. This *writer's block* is a well-known phenomenon and we all will experience it from time to time.

A useful exercise to help you overcome the writer's block consists in *talking* about your research. Indeed, this can be so helpful to writing that some of my colleagues advise their students to give a seminar as a way to practise the art of making clear to themselves and to others the *what* and the *why* of their research project, before they *write* about it.

In my writing workshops, I give students a 10-minute task that can be equally effective. The task is simple: working in pairs, each student takes turn to tell the other *what* their project is about and *why* it is important. The listener is allowed to ask questions at any point and to take notes if (s)he wishes. After ten minutes they swap role. As soon as I start the timer, it is fascinating to see how quickly the room fills with a lively buzz! Invariably, once the exercise is over I ask my students whether they noticed that something amazing had just happened: at my 'go!' they all started talking without the slightest hesitation!

This exercise not only makes it easier for them to approach writing without feeling daunted by the writer's block, it also makes them aware of their audience and the background knowledge they possess. Indeed, the more removed the background of the listener is from the research topic of the speaker, the more the listener will have to ask questions that solicit additional explanations: for example, to clarify the meaning of any jargon or technical terms; to provide additional background information about the specific research area; to place the research in a wider context; and so on. While such questions help the listener to understand, they are also useful to the speaker to further clarify concepts in their minds and to improve on their exposition.

Indeed, after spending months focussing on our own project, we become so familiar with our research and its key aspects that we tend to assume everyone else knows about them too. But this is far from being the case! Thus, when writing, you need to adapt your language to the knowledge that your reader already possesses. Often, minor adjustments can make a major difference in improving the reader's experience. For example, you can include a brief explanation of technical terms in a footnote, add background sections that summarise key aspects of the broader research field, and include well-crafted diagrams or sketches that effectively illustrate the concepts you want to convey (we shall see in Chapter 7 how to create clear figures and tables that further enhance your writing).

3.3 GETTING THE STRUCTURE RIGHT

After you have spent some time talking about your research and clarified its key message in your mind, it is time to prepare to write. The first and most important aspect to get right at this point is planning a proper structure for your content. Indeed, getting the structure right is crucial for clear and effective writing, regardless of the type of document you are writing.

A proper structure provides the skeleton that holds everything together: get it wrong and your entire writing will collapse, no matter how good your research; get it right and see how it enhances the brilliance of your work.

A well laid-out structure is meant to provide a *map* for your reader to follow but it will also make it easier for you to build a story around your project as you write. Like every story worth telling, it must have a beginning (the *what* and *why* of your project), a middle (the *how* it was carried out and what results were obtained), and an end (the *so what?* of your study, namely the difference it makes in the bigger picture). You only need to include all the *necessary and sufficient* information required for your reader to follow and understand what you have done, why and how.

Most likely what you will include in your paper or thesis will be far less than all you will have done or learned during your PhD, or in the execution of your project. Yet, like a film director or a story-teller you must leave out any unnecessary information that would just distract your reader from the essential plot. Keep this simple idea in your mind and see how more effortless and enjoyable the writing process becomes.

3.4 MIND MAPS

An excellent way to prepare the structure of your thesis or paper is by creating a *mind map*. A mind map is a graphical diagram used to organise information in a way that shows a connection of ideas, topics, or items to a central concept.

The usefulness of a mind map rests on the recognition that our brains rarely operate in a linear fashion but are rather involved in a complex pattern of associations and links that can be better reproduced as a 'snapshot' of the mind. For this reason, mind maps are often used by educators, psychologists and others for activities associated with learning, brainstorming, memorising, visual thinking, and problem solving.

When it comes to writing a thesis or research paper, mind maps can be extremely effective because they:

- provide a starting point for planning the overall structure of your document and of its individual chapters or sections;

- help you to organise a large quantity of information without the risk of forgetting key elements;

- can be expanded to a greater level of detail by creating new mind maps for individual components of your document;

- can be easily modified as the project evolves.

An example of a mind map produced for a literature review is shown in Figure 3.1. Here, the central theme was the investigation of a possible relationship between different types of welfare states and gender equality in employment. A number of other key nodes branch out from the central theme and each further develops into additional sub-topics and concepts.

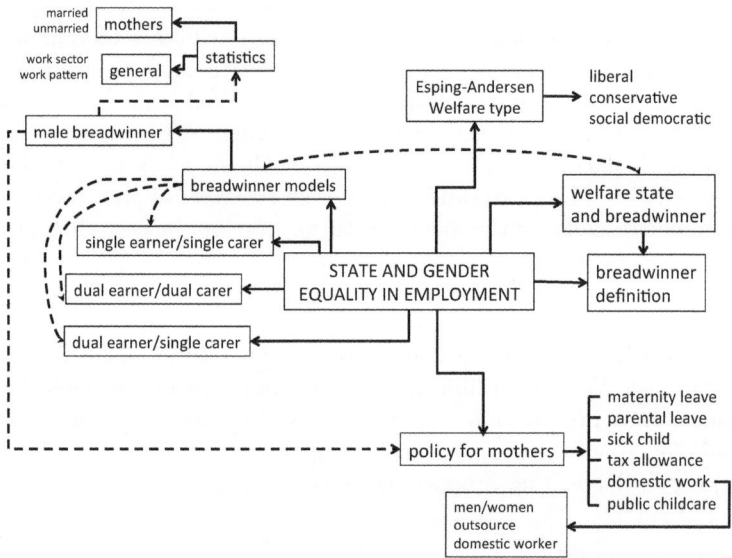

Figure 3.1 Example of a mind map showing a central topic and its various branches with related items connected by lines.

In a similar way, you can produce a mind map (either by hand or using any of the available software on the internet) starting from a core topic at the center of a page and then creating new nodes, each representing key aspects of the central topic.

If you are using a mind map to produce the general layout of your thesis or paper, the core topic will be the thesis itself. Alternatively, you can use the same approach for a specific chapter or section of your thesis or paper. From this central node, start drawing a line (just like the branch of a tree) and at its end write one of the topics that you want to include in this particular chapter or section. Hence, keep branching out to other items that you wish to include, drawing one branch per item. If appropriate, draw lines that link items together (even if across different nodes) to show any potential connection between them.

Once you have exhausted all the topics for a particular section, move back to the main node (your chapter) and draw new branches and sub-branches until you have included all the key elements that will be part of a given node (each representing a chapter or section).

At this stage, you should not be overly concerned if your mind map looks rather messy. That's fine! That is how it is supposed to be at first. Remember, a mind map is a visual representation of what you have in your mind about all the items that you wish to include/present/discuss in your document, chapter or section. So, until you form a clearer idea of how to structure your work, it is not surprising that your first map looks rather untidy.

Once all key items have found a place in your map, you can proceed with a first revision by re-arranging topics (nodes or items) that belong together, leaving out any un-necessary branches, and organising the information in a more logical or sequential arrangement. A possible revised version of the mind map shown in Figure 3.1 is given below (Figure 3.2).

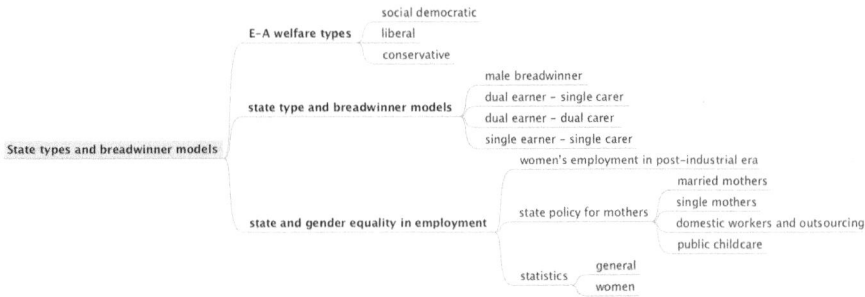

Figure 3.2 Same map as the one in Figure 3.1 but with all its elements arranged by association.

As things keep shaping up in your mind (and in your mind map), you can finally convert your mind map into a *structured layout* of your entire document like the one in Figure 3.3.

Keep revising your layout until you feel the sequence of its various components (chapters or sections) makes logical sense. Specifically, think about what your reader will need to read first in order to be able to understand what follows and use that information as a guide to decide on the right order in which content will be presented.

Use informative titles for chapters, sections and sub-sections that clearly give an indication as to what the reader will find in them, or indeed as to what you are supposed to write in each! As your writing progresses, you should constantly check that the originally intended structure still makes sense and be prepared to revise it if necessary.

As an added benefit of having created a structured layout early on in the writing process, you will be able to decide how many pages you should

Literature Review

Part I: State and gender equality in employment (TOTAL = 10-12 pages)

1 **Types of welfare states (2 pages)**
 1.1 Liberal
 1.2 Conservative
 1.3 Social democratic

2 **Breadwinner models (3-4 pages)**
 2.1 Definition
 2.2 Male Breadwinner
 2.3 Dual earner – single carer
 2.4 Dual earner – dual carer
 2.5 Single earner – single carer

3 **State and gender equality in employment (6 pages)**
 3.1 Women's employment in post-industrial era
 3.2 State policy for mothers
  3.2.1 Married mothers
  3.2.2 Single mothers
  3.2.3 Domestic workers and outsourcing
  3.2.4 Public childcare
 3.3 Statistics
  3.3.1 General: work sectors and work patterns
  3.3.2 Women: married and lone mothers

Figure 3.3 Example of a structured layout from the mind maps of Figures 3.1 and 3.2. Producing a structured layout also gives you a good idea of how much text to write in each chapter (section) based on the estimated total length of your thesis (paper).

realistically devote to each section or chapter. This is especially important for lengthy documents such as a PhD thesis in order to avoid the risk of writing an encyclopaedic tome. Conversely, for research papers with a strict page limit (as for *Letters* or *Rapid Communications*), you will have a better idea of which details to include and which to leave out.

Now that you have an overall structure for your thesis or paper, you can finally begin to write in the knowledge that all the information you have in mind about everything you did in our project has a clear place where to go. It is a bit like having a drawer for everything, or a series of boxes each intended for a specific room in your house. The best strategy to begin writing is then to simply focus on one 'room' or one 'box' at a time, namely a very specific chapter or section of your thesis or paper and "forget" about all others.

This strategy will help you to focus on one thing at a time and to remove any sense of overwhelming that you might experience if you tried to work at the entire document all at once.

3.5 CORE DUMP

If you are anything like me and tend to be a perfectionist by nature, beginning to write may result in what some people call 'zig-zag' writing: you write a sentence, read it, decide you do not like it, delete it. Then you re-write it, re-read it and still decide you do not like it. So you delete it again and try to re-write it; the process goes on and on...

I know this pattern all too well! It is frustrating and highly inefficient as you may end up spending hours to produce just a couple of sentences. In addition, you may also run the risk of forgetting key ideas as you try to polish your text at a stage where you should rather focus on making sure you include all the relevant information that goes in the specific section you are writing.

A much better strategy, at least until you become a more skilled academic writer, consists in 'dumping' your thoughts on paper (better even on your computer) without worrying (not yet at least!) about structure, grammar, punctuation, logical sequence. Just focus on creating content. As we shall see in later chapters, you will have plenty of time to refine and polish the text you have created.

For the time being, simply focus on practising a *core dump*, namely producing a 'large' amount of text in a short period of time. Key to achieving this is to start writing without stopping and, more importantly, without re-reading what you have just written. In order to do so effectively, you need to resist the urge to write something perfect the first time round and simply focus on jotting down the ideas and content as they come to your mind. As an extra trick, consider minimising the editor window on your computer to just one or two lines so that you cannot re-read what you have just written.

My advice would be to do the core-dump exercise for at most 15 minutes at a time. You will be surprised by how much content you can create in such a short span of time if you follow the instructions for this task: Do not re-read what you write, do not worry about clarity or grammar. Just write.

Before you begin, however, make sure you clear your mind from other thoughts and worries that may hinder you during the 'core dump'. Also, switch off any source of distraction: no Facebook, no Twitter, no email alerts. For 15 minutes! Then, open a document editor and start writing. When you finish, take a short break: go for a walk, have a cup of tea, or do anything that distracts your mind for a while. Once you feel rested again, have another 15-minute session of core dump. You may notice that the second time round you are more concentrated and that you manage to write more fluently and with less effort. Just take mental note of any difference at the end of your writing sessions.

If you wish, count the number of words you have managed to write. Later on, you will be able to assess how much of your original text will be retained in the final version, which in itself will provide helpful insight on how long it takes you to go from an initial draft to a final polished version.

The core-dump exercise will become easier to do as you keep practising and within a few sessions, you will have managed to write a good amount of text. In the following two chapters, we shall see how to revise and edit your drafts. This is where you need to spend most of your time.

Chapter 3: The Drafting Step
In a nutshell...

- Get clarity about the message you want to convey. Explain to colleagues and peers what your project is about and why it is important.

- Decide (or find out) who your target readership is and assess the level of background knowledge they already possess.

- Develop a mind map of your entire project by including all the key broad components. Then, for each component (chapter, section, sub-section) develop a new mind map focusing on all relevant items and ideas to be included and revise until all important information is added.

- Using your mind map, prepare a structured layout for the entire paper or thesis or for each individual section or chapter.

- Once the overall structure is in place and you know what content to include, start drafting one section at a time.

- Write as quickly as possible to fix ideas on paper. You can revise and edit your drafts later on.

EXERCISES

3.1 **Think about your audience.** Take an A4 sheet of paper and draw a vertical line in the middle. On the left-hand side, list all different types of audiences your writing will address (e.g., PhD examiner, other fellow students, expert in the field, lay person, ...). Next to each entry, list on the right-hand-side the assumed background for each type (e.g., familiar with broad picture but unaware of specific issues, unfamiliar with formalism, ...). Discuss your notes with your supervisor to decide the level of detail and background that you need to include in your writing.

3.2 **Talk about your research** [10 minutes]. Pair up with a fellow student and take turn to explain to each other *what* your project is about and *why* it is important. If this is not possible, record yourself! It will feel terribly awkward at first, but it is an excellent way to get you going. And don't worry: no-one needs to listen to what you recorded apart from you.

3.3 **Give a seminar about your research** [40 minutes]. Volunteer to give a seminar about your research to different audiences (your research group, your wider department, other scholars at national or international conferences) and see how this affects your presentation.

3.4 **Develop a mind map** [20 minutes]. Take a large sheet of paper and use it in landscape orientation so as to optimise the use of space. Starting from the centre of the sheet, write down a key word representing the main topic of your mind map. Hence, branch out to various sub-topics as discussed in Section 3.4. Once you have included all the relevant items, move on to the next exercise.

3.5 **Revise your mind map** [20 minutes]. Group together items and branches that are related to one another. Use as many sheets as you need until you are satisfied with the outcome. Finally, number each main branch in order to arrange all topics in an ordered, logical sequence.

3.6 **Prepare your layout** [20 minutes]. Using the revised mind map created in the previous task, you can now proceed turning it into a proper layout. Repeat tasks 1-3 for your thesis, your research paper, individual chapters, sections and subsections. Hence, decide roughly how many pages you need to write for each part.

3.7 **Core dump** [2x 15 minutes]. Decide on a specific section you wish to write. Set the timer to 15 minutes: have a core dump, take a break for 15 minutes, have another 15-min core-dump session. Count how many words you have produced in each section. Use this technique a number of times until you have included all the relevant content and completed writing up the whole section. This material will form your first, early draft. We shall see in the next chapters what to do with it.

FURTHER READING

The Writing Center - University of Wisconsin-Madison
  https://writing.wisc.edu/index.html

On mind maps and how to create one
  http://www.tonybuzan.com/about/mind-mapping/
  http://www.mindmapping.com

More on getting the structure right
  https://marialuisaaliotta.wordpress.com/tag/mind-map-to-layout/
  https://marialuisaaliotta.wordpress.com/2012/08/21/city-maps-and-theses-layouts/

CHAPTER 4

The Revising Step

CONTENTS

4.1	The Triage Approach	52
4.2	Common Problems	52
	4.2.1 Faulty organisation	52
	4.2.2 Lack of clarity	53
	4.2.3 Inappropriate usage of language	55
	4.2.4 Poor grammar	56
4.3	Tips for a better structure	56
4.4	Paragraphs as building blocks	58
4.5	Reverse outlining	59
4.6	Linking paragraphs together	60
4.7	Parallel Structure	61
4.8	Feedback: When, What, and Whom to ask	63

MOST STUDENTS tend to assume that once they have finished drafting their chapters, their work is almost done. Not so. In fact, the most time in the writing process should be spent *revising* and *editing*. While the difference between revising and editing is not necessarily clear cut and the two activities largely overlap, the two stages are separated in the context of this book. Here, the revision stage will focus primarily on constantly checking and refining the structure of your drafts. The editing stage, by contrast, will involve polishing the language by making sure that the text reads well and every single word counts.

As illustrated in this chapter, revision is all about organising ideas in a way that makes the text flow according to a logical argument. Occasionally, you may need to move paragraphs around or create new sentences to provide a link between paragraphs.

Revision typically proceeds from the broader items (chapters, sections) through to paragraphs and finally to individual sentences, without getting lost in details until the overall text is properly structured. As we shall see, this implies first and foremost identifying the most pressing issues and dealing with each in order of importance.

4.1 THE TRIAGE APPROACH

The first, most important aspect of revision is to assess the main issues of your early drafts. One of the best pieces of advice on what to do with them is presented in the excellent book by Patricia Goodson, *Becoming an Academic Writer – 50 Exercises for Paced, Productive, and Powerful Writing* [2]. She suggests to revisit your text as if *triaging* a wounded patient.

Triage is a very effective approach used by medics to determine the priority of emergency treatment for large numbers of patients or casualties. Those of you who have had the misfortune of being taken to an emergency department will know that the very first thing doctors do is to assess the severity of your condition. Is it so serious to be life threatening? Does it require prompt attention, but there is no danger for your life? Or is it simply an injury that can be treated after more serious patients have been dealt with?

Similarly, your early drafts will likely show a number of issues that need attention. However, not all of them have the same level of urgency and while some must be dealt with first, others (typically grammar mistakes or typos) can wait for a later revision before being tackled. In the following section, we look at the most pressing problems of early drafts and the order in which they should be addressed.

It is especially important to tackle each issue in turn rather than attempting to solve all in one go. This is because unless the main problems have been addressed, there is little point in focussing on the details. If the main problem of your text is a wrong organization of material, you need not obsess about the style of a specific sentence only to realise later that in fact you do not need that sentence at all!

4.2 COMMON PROBLEMS

According to [2], and also in my experience, the most pressing issues of early drafts fall in one or more of four main categories, namely (in order of severity):

- faulty organisation;
- lack of clarity;
- inappropriate usage of language; and
- poor grammar.

These are briefly addressed in the following sections.

4.2.1 Faulty organisation

Faulty organisation is normally associated with poor structure and is perhaps the most common problem of early drafts, especially for inexperienced writers. Even if you have spent time preparing the overall layout of your chapter, chances are that the text you produce will likely require additional checking. This may entail moving paragraphs around to ensure that the text flows in a logical way; creating new material to provide a transition between paragraphs; or most likely a combination of both. Since getting the structure right is such an integral part of the revision stage, most of the rest of this chapter will deal with ways of checking and improving the overall structure of your text.

4.2.2 Lack of clarity

Lack of clarity may be due to a number of different factors, including poor structure. However, even if the structure is fine, clarity gets compromised when the concordance between certain grammatical parts of a sentence is lost, as with *dangling participles* (Section 6.2.4) and *ambiguous referencing* (Section 6.2.5). Sometimes, also the use of awkward language may confuse your reader and make your text unclear. These and other stylistic issues will be discussed in detail in the next chapter.

An example of an excerpt whose main issue is lack of clarity is shown below, from an early draft by a PhD student of mine.

Excerpt from an early draft

A possible alternative mechanism from the proton scattering can be achieved through the use of a gas target in place of a solid CH_2 target... While the results of this experiment are yet been published and the context of the experiment differs to that of this work, the methodology offers an alternative means for study of proton scattering studies. As the experiment uses a beam which is an isobar of ^{21}Na, these results should also provide data on scattering yield, which will provide a good indicator to the feasibility of conducting proton scattering with ^{21}Na.

Even if you are familiar with the topic – and chances are you are not! – you would still struggle to understand what the student is trying to say. So, let me give you a little bit of background. We had studied a nuclear physics reaction that involved accelerating a beam of ^{21}Na ions onto a solid target (a thin plastic CH_2 foil) containing hydrogen (i.e., protons). Specifically, we were interested in the so-called resonant elastic scattering between protons and ^{21}Na (it really does not matter if you do not understand the topic). Soon after we completed our study, we found out about a similar investigation by a group of colleagues who were using a similar approach, except with a slightly different beam of particles (**an isobar of** ^{21}Na) onto a *gas* (rather than solid) hydrogen target. Their study was still ongoing and no results had been published yet. However, if successful, their approach would have provided an alternative means of investigating the same reaction we were interested in.

If you now re-read the excerpt in light of this background information you may start making more sense of it, but most likely you will still find the text unclear. In this example, the most pressing issue of the excerpt is not so much its structure, but rather the ambiguous referencing, mostly due to poor grammar and repetitious language. Let's see how.

The terms **this experiment** and **this work** in the second sentence clearly refer to two different experiments (the first instance refers to the experiment performed by our colleagues; the second to our own experiment). Yet, the use of the demonstrative article **this** for both experiments confuses the reader. In the same sentence, the phrase **are yet been published** is grammatically wrong and therefore unclear. The correct phrase would either be **are yet to be published** or **have not been published yet**. At the end of the same sentence, **for study of proton scattering studies** does not necessarily compromise clarity, but makes an odd use of language as one does not **study... studies**.

Further on, **As the experiment uses a beam...** does not make it clear which experiment we are talking about: is it ours or that of our colleagues? Note that the ambiguity here is due to the use of the determinative article **the** without further qualification. Similarly, does **these results** refer to those obtained in their work or in ours (i.e., the one presented in my student's thesis)? Last but not least, repetitions such as **An alternative mechanism... offers an alternative means** and **these results should provide... which provide...** should also be avoided.

I should perhaps point out that, when writing, we obviously have clear in our mind the thought processes (and background knowledge) that lead us to write in the way we do. So, for example, the student author of this excerpt obviously thought that his text was clear. However, as you have just realised, this is not necessarily the case for your readers. In other words, your readers do not benefit from the chain of thoughts and background information that you have, unless you make both explicit to them. Thus, as a writer, your main task it to anticipate what your readers may find difficult to follow and give them the information they need.

Whenever your text lacks clarity, it can be useful to answer the following questions:

- What do you want to say?

- Are you saying it?

- What do you want your reader to remember?

And then simply answer each question in the clearest possible way, by focussing on one key idea at a time. For the example just discussed, answering these questions could lead to a bullet list like the one in the following box.

> ### EXCERPT FROM AN EARLY DRAFT (REVISED)
>
> - An alternative approach for proton elastic scattering studies consists in using a hydrogen gas target in place of a solid CH_2 target.
> - An experiment, using an isobar of ^{21}Na on a H_2 gas target, is currently underway and no data have been published yet.
> - If successful, the new study will provide useful insights on the possibility to adopt this approach to investigate proton scattering with a ^{21}Na beam.

Now you can simply remove the bullet points and retain the text as is. I hope you will agree this is much clearer than the original draft.

4.2.3 Inappropriate usage of language

As we have seen (Section 3.1), being aware of the audience you are writing for is critical in academic writing. This typically requires using a language that can be understood by your readers and involves providing a sufficient level of detail and background information. However, while the type of information presented is dictated by the type of audience, the level of formality, also called *register*, is a defining feature of all academic writing. In other words, the language used in academic writing differs from the language we normally use, for example, when discussing with colleagues. In particular, academic writing typically avoids the use of:

- colloquialism or excessive jargon;
- the second personal pronoun *you* (though the use of personal pronouns in the first person may sometimes be allowed, as discussed in Section 5.3.4);
- phrasal verbs (Section 10.1.9), especially when other more formal alternatives exist (for example, **to investigate** rather than **to look into**);
- redundancy and ambiguity;
- contractions (**do not** rather than **don't**);
- imperatives, except in phrases such as **for further details, refer to** (Smith, 2016) or **see Figure 3.4**.

On the other hand, formality does not necessarily mean having to write in an overly abstract and inaccessible way. So, for example, the sentence ([14], p. 101):

Developing regular exercise programs and diet regimes contributes to disease risk prevention and optimal health promotion.

becomes much clearer and direct by avoiding unnecessary sophistication and redundancy as:

Regular exercise and attention to diet help prevent disease and promote health.

As we shall see in greater detail in Chapter 5, scientific language possesses additional features as it is also factual, objective, non-emotional, and normally cautious. The *tone*, namely the writer's attitude towards himself, his subject and his audience, is often neutral in scientific writing, unless expertly used (for example, in grant applications) to portray excitement and confidence rather than hesitation and doubt, through a proper use of language and the careful selection of words. Examples of different types of register and tone are presented in the excellent book *Writing Science in Plain English* ([14], p. 7-11) by Anne Greene.

4.2.4 Poor grammar

Grammar is an essential part of any language. It establishes a set of rules on how words and sentences are combined together to provide meaning. Without grammar, languages would reduce to a mere collection of sounds and high-level communication would be impossible. As such, grammar provides the structure that holds sentences together in much the same way as the skeleton supports the rest of a body. A good knowledge of grammar is essential for good writing, whether academic or not.

Obviously, native speakers of English instinctively know what is right and what is wrong even though many of them are blissfully unaware of some of the most basic grammatical rules of their own language. For non-native speakers, however, achieving a good knowledge of English represents an extra hurdle to overcome to be able to master the art of effective communication (whether written or spoken).

It should not surprise then that a common problem with early drafts resides with poor grammar. In fact, I was once mentoring a student whose English was so poor that it became virtually impossible for me to help him as I struggled to make sense of what he was trying to say.

Clearly, enlisting the help of a native speaker to check and correct your grammar is no substitute for putting effort into improving your knowledge in the longer run. And yet, you can still make significant progress by familiarising yourself with some basic elements of English grammar. These are grouped together in Chapter 10 so you can easily refer to them whether you are a native speaker or not.

4.3   TIPS FOR A BETTER STRUCTURE

In the previous chapter, we have already discussed how to organise the high-level structure of a thesis (or paper) and its various parts (chapters and sections) by developing an initial mind map into a structured layout. Once the overall structure is in place, you should approach writing each chapter and section by following one of several possible structures [14] as illustrated below.

The *pyramidal structure* is one that proceeds from the general to the specific, progressively moving towards a greater level of detail. This type of structure allows you to present the broader context first so that your reader can anticipate and make sense of any more detailed information that follows. In its 'general-to-specific' pattern, the pyramidal structure is often used in introductions to gradually guide the reader towards the topical focus presented in the study. In its inverted 'specific-to-general' pattern, it is normally used in conclusion sections, where the focus moves away from the specific results obtained in the study onto their impact in the broader context.

A similar pyramidal structure can be used when proceeding from the most important to the least important points, or viceversa. Because readers remember best either the beginning or the ending of a chapter, a section, or a paragraph, you should exploit these positions for added emphasis or for the most important points that you wish to convey.

The *logical sequence*, as the name suggests, is a type of structure that presents information that is logical in argument, such as: If 'statement A', then 'statement B' or 'Statement A; therefore, 'Statement B'. It is often used, though not exclusively, in mathematical or theoretical work to demonstrate a thesis or proposition as an outcome of given hypotheses and lines of thinking.

A *chronological* structure is useful when describing methods or procedures, especially if a strict temporal sequence is part of the approach used. Occasionally, it can also be used to present results, for example, of a number of closely related experiments or measurements. In such cases, a temporal structure will help in achieving overall order and cohesion. However, if the results from an initial method have lead to a new direction of investigation, a logical structure (rather than a chronological one) may be more appropriate to present a coherent story.

The *compare and contrast* structure is often encountered in the literature review where a critical appraisal of the work of others should be presented. The same structure can also be used in the discussion section when comparing results from our study to those obtained by others.

Finally, the *problem to solution* structure is often, though not exclusively, found in introductions to justify the purpose of the work undertaken and is especially appropriate in grant applications as a way to persuade the readers (and reviewers!) of the importance of the research you propose.

Regardless of the specific type of structure you use in various chapters or sections, you will eventually need to coordinate them all together by reminding your readers on where they are at each point along the way and of where you intend to take them next.

In my workshops, I often illustrate this concept by using an analogy with city maps. When you visit a new city, the first thing you likely do is to get a map and locate yourself in it to see where you are in relation to the rest of the city. If you then wish to go to a specific place, you decide on the best path to take you there and possibly identify some landmarks along your way that will allow you to check if you are progressing in the right direction.

In a similar way, you need to provide a map to orientate your readers on where they are (the broader context), where you intend to take them (the final purpose of your project) and how to finally get there (the *what* was done and the *how*).

Once you are happy with the overall intended structure of individual chapters and sections, you can move on to tackle the structure of individual paragraphs and that of sentences.

4.4 PARAGRAPHS AS BUILDING BLOCKS

Paragraphs fulfil two main purposes: 1) to break down the text into visual chunks for easier reading; and 2) to provide the building blocks needed to construct your text.

Like chapters and sections, paragraphs too have a well-defined structure [14], typically consisting of an *opening* statement(s) to introduce the reader to the main topic or idea of the paragraph, a *development* part to add information or supporting statements and to build arguments towards a conclusion, and a *conclusion* that summarises the information presented. Most paragraphs will also contain transition elements to connect paragraphs together.

PARAGRAPHS AS BUILDING BLOCK

(Transition) + Key Idea + Development of Key Idea

An example of a well-structured paragraph is shown below.

EXAMPLE OF A WELL-STRUCTURED PARAGRAPH

A paragraph should *also* be consistent in structure; that is, it should complete three functions in order. *First*, the paragraph should open with a transition. The transition can be as short as a word or a phrase that was used in the previous paragraph – or as long as a sentence or even two or three... *Second*, the transition should be followed by a topic or key sentence. *Third*, the rest of the paragraph should provide support or evidence for the idea in the key sentence. As you revise, you should check each paragraph against this template.

From: Gray, 2005 [15], p. 42 – cited in Goodson, 2013 [2], p. 99, and Gray, 2015 [16], pp. 47-48.

The paragraph is well structured because it provides one key idea (**A paragraph... should complete three functions**) supported by additional sentences that develop the key idea. These are the sentences introduced by **First**, **Second**, and **Third**, where additional information is provided to further support what stated in the main or key idea. Notice how the presence of a transition word (**also**) provides a link with what will have been said in a previous paragraph. No new concepts or ideas are presented in this paragraph.

When a paragraph is well structured, the readers know what to expect and can easily make sense of the content even if they may miss out on details. Different is the case for a paragraph that lacks structure, like the one shown below.

EXAMPLE OF A POORLY STRUCTURED PARAGRAPH

A paragraph should also be consistent in structure; that is, it should complete three functions in order. First, the paragraph should open with a transition. Transitions are essential for connecting the current paragraph to the previous one(s), and without transitions authors may lose their readers as the paragraphs begin to read like bulleted lists of ideas. Bulleted lists might be very useful in certain parts of the text, but they aren't fully developed text. Many readers don't appreciate bulleted lists.

From: Goodson, 2013 [2], p. 99.

In this case the key idea has not been fully developed. Instead, new ideas are presented, which however take the reader off by the tangent and leave him/her with a sense of confusion and loss of direction.

Early drafts do not typically contain well-structured paragraphs. It is important then to focus first on revising each paragraph so that only one or two key ideas at most are presented in each.

4.5   REVERSE OUTLINING

Once you have checked the structure of individual paragraphs by making sure that each contains one or two key ideas at most, you can proceed to verify that paragraphs are presented in the right order to the reader, thus effectively double-checking the overall structure of the entire section or chapter.

One way of doing so is by using *reverse outlining*, a term I first encountered in *Explorations of Style*, a blog about Academic Writing by Rachel Cayley (you can find the link to the blog post in the Further Reading section). The process briefly consists in extracting the key idea(s) from each paragraph and copying them into a separate file in the same order in which key ideas appear in the text.

By focussing your attention on the key ideas alone, you are now in a better position to check their logical sequence rather than being distracted by unnecessary details. This allows you to see if any logical link is missing or whether some key ideas (and therefore their associated paragraphs) need to be moved around. If so, you can first establish the right order of all key ideas in the separate file, and then re-order accordingly all the relevant paragraphs in the main text.

When all paragraphs have been re-ordered in the right logical sequence, just make sure that transitions between them are appropriate and check for overall cohesion.

4.6 LINKING PARAGRAPHS TOGETHER

When linking paragraphs together, transition words such as *however, nevertheless, moreover, in addition*, are often used to great effect as they signal to the reader a change of focus or emphasis.

In fact, you can highlight different types of structures in your paragraphs, by using appropriate transition words. For example [14], transition words such as *first, second, meanwhile, finally* clearly indicate a chronological structure; *for example, specifically, to illustrate* may be used to indicate a general-to-specific structure; *most importantly, the foremost, clearly* for a least-to-most-important structure; *similarly, by contrast, in comparison* for a compare-and-contrast structure; *but, however, nevertheless, instead* for a problem-to-solution approach.

Overusing transition words, however, often indicates problems with the underlying structure of your text. If that is the case, you should revise the structure first rather than searching for yet another transition word.

The next example [17] beautifully illustrates how different paragraphs are connected with one another by making appropriate use of transition sentences that overall preserve a sense of structure and direction (the key ideas in each paragraph are shown in italics).

> LINKING PARAGRAPHS TOGETHER
>
> *The core sentence will often be the opening sentence of the paragraph.* In this position, the sentence will function as a sort of title, which tells the reader right away what the paragraph is about. This way, the reader will be better prepared for the other sentences in the paragraph, which contain complementary information and find it easier to skim through the text.
>
> Some paragraphs have a different structure. They open with a number of introductory remarks which prepare the reader for the core idea, or which guide the reader from specific details to a more general statement. For instance, the writer may start off with a number of facts or observations, and extrapolate from them. *In such paragraphs the core idea can be found in the last sentence.*
>
> We have noted that ideally the core sentence will come at the beginning of a paragraph. But that doesn't mean that it needs to be the very first sentence. *Sometimes the core sentence will be the second or even the third sentence.* In the first sentence (or the first few sentences) the writer will make a link with the previous paragraph(s). Experienced readers will recognise this link, and know that they can expect the core sentence soon after.
>
> *Sometimes a paragraph will have two core sentences.* That happens typically when a paragraph opens with a particular statement, followed by a number of arguments. When these arguments take up a lot of space in the text, it may be useful to repeat the statement after the arguments which support it. This will make it easier for the reader to keep track of what is being argued. *Such a paragraph will therefore have two core sentences: one at the beginning and one at the end.*
>
> From: *Effective Scientific Writing*, Marie Curie Doctoral Training - Course Handbook by M. Caenepeel, 2012 [17].

4.7   PARALLEL STRUCTURE

The next structural check can now proceed at the level of individual sentences. The simplest structure for a sentence consisting in a single main clause[1] is one in which subject and verb are close together, either followed by an object or not. For complex sentences that contain a main clause and one or more

---

[1] If you are not familiar with the grammatical terms used, please refer to Chapter 10.

subordinate ones, the simplest structure is one in which the main clause comes first and the subordinate one(s) follows.

However, a text in which all sentences are constructed in the same way, have a similar length, and always follow the same pattern (main+subordinate, or subordinate+main) is often dull and boring to read. Far better is to vary both the structure, the pattern and the length of sentences and use these elements for added emphasis or improved style.

There is one situation, though, where a specific sentence structure should be adhered to. If a sentence presents several pieces of information in the form of a list, the sentence must follow a *parallel structure*. Parallel structure refers to the grammatical form of the items in a list. These can be nouns, phrases or clauses, each containing similar elements, but the form chosen for the first item on the list must be repeated in all others.

The following examples illustrate the concept:

Their *careful planning* (adjective + noun) and *proper execution* (adjective + noun) resulted in a successful expedition.

He failed the exam not only *because of poor preparation* (prepositional phrase), but also *because of bad time management* (prepositional phrase).

Objectivity in science resides in accepting *what we observe* (objective clause) and not *what we believe* (objective clause).

A lack of parallelism can obfuscate the meaning and make sentences awkward to read, as shown in the following example [18]:

Authors of academic papers should avoid [1] an informal style, [2] unnecessary wordiness, and [3] writing a poorly organised paper.

Here, the three items in the list do not have the same grammatical form, namely they break parallelism and obfuscate clarity. A way to address lack of parallelism consists in listing all items in turn and identifying their grammatical form. Specifically, in the example above:

[1] is a direct object (adjective + noun)

[2] is a direct object (adjective + noun)

[3] is an objective clause (here introduced by the verb **writing**.)

Rephrasing a sentence that lacks parallelism then requires to select a specific grammatical form for the first item and maintain it throughout for all others (for example, all nouns or all adjectives) as in, respectively:

Authors of academic papers should avoid [1] an informal style, [2] unnecessary

wordiness, and [3] poor organization.

or, better even:

Academic papers should be [1] formal, [2] concise, and [3] well organised.

4.8 FEEDBACK: WHEN, WHAT, AND WHOM TO ASK

Having completed a first revision of your early drafts by ensuring the overall structure is coherent and reflected in a proper sequence of paragraphs and sections, it is time to ask for feedback.

Getting (and acting on) prompt, useful and constructive feedback on our drafts is the single most important activity that will improve your writing. Sadly, this is often one of the worst experiences for students. I often hear students lamenting that they either get little or generic feedback from their supervisor (something like *"it's not good enough, you need to rewrite this section"*); that the feedback they get is just too late to be of any use (they have already drafted another three chapters in the meantime!); or that they get so much and such detailed feedback that they feel overwhelmed and get stuck!

As supervisors tend to be very busy people, a better approach to getting feedback consists in being strategic about *whom* to ask and specific about *what* to ask for: both will depend on how advanced your draft is. Again, some excellent advice can be found in *Becoming an Academic Writer* [2] and is summarised in Table 4.1.

If you are working at an early draft, chances are you are still piecing together the content and deciding what you should include in your text and what you can safely leave out. Thus, the type of feedback you mostly need is on: whether the content is appropriate (i.e., does it match the expected background knowledge of your readership); whether it is enough (i.e., are you providing enough detail for your readers to understand what you are trying to say?); and whether it is relevant (i.e., are you presenting only necessary and sufficient information for your readers to understand, or are you desperately trying to cram into your thesis all you have done or read about?).

In such cases, you do not necessarily need to involve you supervisor, but can still get valuable feedback from a fellow colleague or another student. In fact, unlike your supervisor, non-experts in your field might even be in a better position to tell you whether you need to include additional background information, for example something that you assume you readers already know whereas in fact they do not.

Once your draft reaches a middle stage of revision, your focus needs to shift to its overall structure and style, so you can decide whether the ideas you are presenting are clearly explained and form a coherent, well-structured piece. Here, a more experienced colleague (including you supervisor) can offer a critical eye and help you assess any potential weakness.

Table 4.1 Overview table on *when*, *what* and *whom* to ask for feedback on your writing.

When to ask	What to ask for	Whom to ask
Early Draft	Focus on the *'What'*: Are ideas appropriate, enough, relevant?	Fellow colleagues and students
Middle Draft	Focus on the *'How'*: Are ideas clear, coherent, well structured?	Experienced post-doc; your supervisor
Final Draft	Focus on the *'Overall Content'*: Is the work correct, accurate, comprehensive?	Your supervisor; other experts

Once you produce a final draft, you need expert advice to check and comment on whether the information presented is correct, accurate, and comprehensive; whether you have cited all the relevant works; and whether the analysis and results presented are sound.

Unless you follow this strategy, there is a real danger in asking your supervisor for feedback on early drafts without also giving them specific instructions on what feedback you need. The danger is that your supervisor may try to give you feedback on everything simultaneously: on structure, content, style, clarity, accuracy, grammar, punctuation, references, figures, tables... This is simply too much for anyone to take on board! And you certainly do not want to feel overwhelmed, disheartened and stuck!

Similarly, please do not wait until you have written your full thesis or paper before asking for feedback. This is a sure recipe for disaster as you may have to change so much all at once that you might just as well give up!

On this note, I wish to add a final remark. We often tend to take feedback quite personally. I certainly did and I still remember the feeling of inadequacy I experienced when my supervisor returned my first chapter draft full of red ink all over the pages. I felt shattered not just because of the extent of what still needed attention, but because somehow I took that as a reflection on me as a person. And I felt I simply was not good enough! Luckily, with time I have come to realise that when I get feedback on my writing, this is precisely that: a constructive criticism of *my writing* and not of me as a person.

Now, whenever I offer feedback to my students, or whenever I mentor a student who complains about the feedback they receive, I always remind them not to take things personally – even more so, when the feedback is extensive and very detailed.

Think about it this way: if someone (your supervisor or anyone else) has taken the time to read your drafts and has gone to such an extent as to provide you with plenty of advice on how things could be improved, then this is a clear sign that they care about you and want you to succeed! After all, it would be a lot easier for them to just say "fine!" and not to care.

Chapter 4: The Revising Step
In a nutshell...

- Revise your early drafts by tackling the most pressing issues first! Specifically, pay attention to each of the four most common problems of early drafts: faulty organisation, lack of clarity, inappropriate usage of language, and poor grammar.

- Constantly check the underlying structure of your text by making sure that the discourse follows a clear and appropriate pattern.

- Different types of structure are possible and should be used depending on context. The most common ones include: pyramidal, logical, chronological, compare and contrast, from problem to solution.

- Provide your readers with all the necessary and sufficient information to allow them to understand what follows.

- Build paragraphs around one key idea. For each paragraph, provide an opening (often a transition word or phrase), a development, and a closing point.

- Get feedback at each stage of your writing, but be selective and specific about whom to ask and what to ask for.

EXERCISES

4.1 **Build your paragraphs** [20 minutes]. Working on an early draft, group similar ideas together into individual paragraphs (typically one idea per paragraph). Hence, focus on the logical sequence in which paragraphs should appear in your text. If necessary, generate more text to create links and number each paragraph progressively.

4.2 **Build the text layout** [20 minutes]. After completing the previous exercise, take one paragraph in turn and highlight the sentence(s) that contains the key idea. Hence copy and paste each key idea sentence (numbered according to the paragraphs it belongs to) into a new 'Core Ideas' file. Identify the function of each other sentence in the paragraph (transition, connection, development, etc.). If you encounter an 'orphan' sentence, i.e., one that does not fulfil any specific function, pull it out or reword it so that it serves some purpose. Do not skip any sentence.

4.3 **Reverse outline** [30 minutes]. Referring to the 'Core Ideas' file created in the previous exercise, check the order in which information is presented. Does the outline reveal a structure? if not, move sentences around until the sequence is right. If gaps are present, you may need to generate new 'core idea' sentences. Finally, tackle changes in the main text according to the new order of the core ideas outline and check that all transitions and connections in the text make sense.

FURTHER READING

Murray, N. and Hughes, G. (2008) *Writing Up Your University Assignments and Research Projects - A practical handbook.* Open University Press.

Holtom, D. and Fisher, E. (2006) *Enjoy Writing Your Science Thesis or Dissertation!* Imperial College Press.

Tips on academic writing style
   http://homepages.inf.ed.ac.uk/jbednar/writingtips.html

Tips on structure
   https://marialuisaaliotta.wordpress.com/2012/08/21/city-maps-and-theses-layouts/

Reverse outlining
   https://explorationsofstyle.com/2011/02/09/reverse-outlines/

CHAPTER 5

The Editing Step

CONTENTS

5.1	How good is your writing?	68
5.2	Scientific Style in Academic Writing	69
5.3	Pearls of wisdom: Advice for a better style	72
	5.3.1 Verbs in action	73
	5.3.2 Verbs in disguise	74
	5.3.3 Active or passive?	75
	5.3.4 *I* or *we*? Personal pronouns in scientific writing	76
5.4	De-clutter your text	77
	5.4.1 Awkward phrases and waste words	78
	5.4.2 Transition words	78
	5.4.3 Redundant information	79
	5.4.4 Negative statements	79

SOME PEOPLE say that editing is where writing meets elegance. If you wish to make your writing pleasant to the reader, then you should be prepared to spend enough time editing and polishing your drafts. As most readers are easily distracted, your main task as a writer consists in communicating information and knowledge in an interesting and engaging way. This requires not only familiarising yourself with the language used in your discipline, but also learning to use it properly and effectively. As Anne Greene points out in her little gem of a book *Writing Science in Plain English* ([14], p. 2), being able to write well is especially critical in science:

> As science becomes more specialised and the writing more complex, specialists in different fields struggle to understand one another. Poor writing also makes it more difficult to apply discoveries from one field to another, a cross-fertilisation that has advanced scientific discovery in the past.

Indeed, one may argue that unless you can clearly communicate the importance and impact of our research, true scientific progress is either slowed down or hindered altogether, regardless of how good your research is.

This chapter offers specific advice on how to improve your writing style so that your communication becomes clearer and more effective. Since many of the problems of scientific writing are the same regardless of discipline, the principles of good style illustrated here will be valid whether you are a chemist, a biologist, a physicist or a computer scientist.

Finally, note that throughout this chapter, I use several grammatical terms that refer to parts of speech and basic sentence structures. If you are unfamiliar with these terms or are unsure about their meaning, please refer to Chapter 10: *Elements of English Grammar*.

5.1   HOW GOOD IS YOUR WRITING?

In her excellent book, *The Writer's Diet* [19], Helen Sword uses a very apt metaphor to highlight the importance of a good prose. She says ([19], p. 1):

> *Imagine yourself recruiting a long-distance runner to deliver an important message. What kind of person will you choose: a lean, strong athlete with well-toned muscles and powerful lungs, or a podgy, unfit couch potato who will wheeze and pant up the first few hills before collapsing in exhaustion? The answer is obvious. Yet far too many writers send their best ideas out into the world on brittle-boned sentences weighted down with rhetorical flab.*

Sadly, this sounds pretty familiar for most scientific writing I have come across. Luckily there are better ways to communicate our ideas clearly and effectively. However, before we can learn what they are, we need to be aware of the main pitfalls in our own writing and so it is appropriate at this point to test how good our writing already *is*.

This is where the *The Writer's Diet Test* [20], by Helen Sword, can come to help. The test is a diagnostic tool designed to provide a quick reality check on the "fitness" of your prose. All you need to do is to inspect a sample of your text against five key components of any sentence, namely: *verbs*, *nouns*, *prepositions*, *adjectives/adverbs*, and other 'waste words' (it, this, there, due to the fact that, ...). Your prose is then rated against five different 'fitness' categories as [19]:

Lean	as good as it gets
Fit and trim	in excellent condition
Needs toning	would benefit from some attention
Flabby	some revision or editing required
Heart attack territory	major restructuring and editing needed

The test is not meant to be taken overly seriously, but it will provide you with an immediate overview of potential problems in your writing. The test can be performed by hand, by highlighting the occurrence of each of the five grammatical elements mentioned above, or electronically at the www.writersdiet.com website. All you need to do is to copy and paste an excerpt of about 100-300 words of your own prose in the space provided on the

The Test

verbs					
nouns					
prepositions					
adjectives/adverbs					
it, this, that, there					
lean	fit & trim	needs toning	flabby	heart attack	

your diagnosis

FLABBY

test new sample

Your sample has 146 words

The under-representation of women at all stages of physics education is a growing concern. Results from the Institute of Physics (IOP) indicate that the proportion of females entering undergraduate physics has remained consistently lower than for other sciences, with participation rates remaining below 20% during the past 15 years (IOP, 2012). In addition to this, there is indication that a potential gender disparity exists in student performance at undergraduate level (Lorenzo et al., 2006). There is increasing evidence to suggest that specific teaching methodologies and pedagogies have both a measurable effect on students' overall performance and can increase learning more than traditional methods (Hake, 1998). Hake's analysis of the normalized gains on pre- and post-instruction test scores on the Force Concept Inventory (FCI) of nearly 6000 students at US institutions has provided a benchmark for comparison of effective teaching methodologies (Hestenes and Wells, 1992). Learning gains on courses taught to facilitate interactive engagement of students were approximately twice those in traditionally taught courses

Figure 5.1 Sample outcome of the Writer's Test for an early draft excerpt.

website. Ideally, the sample should be representative of your best, polished academic prose.

As soon as you run the test, you will receive an immediate "diagnosis" showing how your writing scored (both overall and in individual categories) on a spectrum of *lean* to *heart-attack*. All elements of your sample belonging to the same category (e.g., verbs, nouns, etc.) will appear colour-coded for easy recognition as shown, albeit in black and white, in Figure 5.1.

As stated by its author, the test has been calibrated against a number of different types of prose, including fiction, journalism, poetry as well as academic writing, and invariably it may be perceived as fairly subjective or perhaps not entirely appropriate to *scientific* academic writing. Nevertheless, it is a good starting point to help you pay attention to your choice of words and how you use them. Once you become more aware of the issues that may adversely affect the quality and readability of your text, you should be in a better position to tackle them and make your prose more pleasant to read.

5.2 SCIENTIFIC STYLE IN ACADEMIC WRITING

As we have seen in previous chapters, academic writing differs from other types of prose in that it is more formal; it uses discipline-specific vocabulary; it is generally impersonal; and it references other people's work. In addition to these features, *scientific* style in academic writing also requires a language that is: *accurate*, *specific*, *concise*, *clear*, *cautious*, and *objective*. Let us see in more detail what these features mean.

Accurate: Words used in a scientific context usually convey very specific meanings that may differ from those of everyday life. For example, *speed* and *velocity* are often used interchangeably in everyday language, but have a slightly different meaning in physics as the former is a scalar quantity and merely refers to a magnitude (for example, 30 km/h), while the latter is a vector and also provides an information of the direction of travel.

Similarly, adjectives such as *efficient* and *effective* are not quite the same even though they are often used interchangeably in everyday language; likewise for *mass* and *weight* which have very different meanings (and units) in physics but are freely used as synonyms in daily life.

Great care should also be used with words that have more than one meaning, as for example **current** which can indicate either something "of the present time" or can refer to a flow of water, air, or electric charges. Even though the meaning might be obvious from the context, it is good practice to restrict the use of technical terms to their very specific meaning to avoid potential ambiguity.

Specific: Terms such as **long, very,** and **high** do not necessarily mean much in science. They are perceived as vague (how long is *long*?) and do not provide the sort of information often required in scientific writing. Always adopt a specific, quantitative language wherever possible. So, for example, rather than saying:

> The simulation ran quickly

be more specific and provide a quantitative statement such as:

> The simulation required only 2 hours on a single computer processor.

Similarly, rather than writing:

> The NaCl concentration in the samples was checked frequently,

state:

> The NaCl concentration in the samples was checked every 15 minutes.

If quantitative language is not possible, use qualifying adverbs or comparative phrases such as, for example:

> The data processing took longer than expected.

Concise: Good scientific language is brief and to the point. Ideally, you should avoid overly long[1] and complex sentences, redundant words that do not add

---

[1] According to Griffies, Perrie and Hull, 2013 [21], the average length of sentences in scientific writing is only about 12-17 words.

meaning, and any unnecessary repetition. Also remember that the active voice shortens the length of sentences and is normally easier to understand. As a rule of thumb, try to use one idea per sentence, and one theme per paragraph.

Clear: Clarity is essential in scientific writing. Anything that leaves the reader with a sense of ambiguity will hamper the effectiveness of the message you intend to convey. Thus, you should strive for both clarity and brevity by avoiding run-on sentences that lead to reader fatigue, as in:

> We ran a model simulation of the setup for the investigation of the dependency of the detector's efficiency on the energy of the gamma rays.

Instead, write:

> We ran a setup simulation to investigate how the detector's efficiency depends on gamma-ray energy.

Sometimes, clarity is compromised by an awkward ordering of words which makes the sentence ambiguous. For example, in the sentence:

> We artificially increased the error bars of the cross section obtained in this work from data acquired by the two detectors placed at $120°$ by a factor of 2

it is not clear whether the data acquired with the two detectors are still part of this work. Similarly, the **by a factor of 2** is simply too far away from the main verb **increased**. A far clearer meaning can be achieved by simply re-ordering the words as:

> We artificially increased by a factor of 2 the error bars of our cross-section data obtained with the two detectors placed at $102°$.

Note that **in this work** has also been replaced by **our** to qualify where the **cross-section data** come from.

Cautious: Very rarely can scientific results be regarded as absolutely certain and definitive unless after having withstood the test of time and experimentation. Even then, as Einstein once famously quoted: *"No amount of experimentation can ever prove me right; a single experiment can prove me wrong"*. It is therefore wiser to avoid strong statements and resort instead to softer, more cautious expressions such as **These results suggest....** Of course, depending on the specific situation, different degrees of certainty can be signalled by the use of appropriate language. A list of useful verbs and adverbs is provided in Table 5.1.

Objective: Communication in science is not about beliefs or personal opinions, but rather about a logical deduction based on (often experimental) objective evidence. As such, subjective language is rarely used and only relegated to cases where individual opinions matter.

Table 5.1 Useful verbs and adverbs indicating various degrees of certainty.

Low certainty	Medium certainty	High certainty
seldom,	probably, perhaps, likely,	certainly, definitely, totally,
rarely,	generally, normally,	always, never, must,
unlikely,	seems to be, suggests,	should, would, absolutely,
improbable.	can, could, may, might.	undoubtedly, impossible.

However, when reporting final results or conclusions of more general validity (beyond just personal opinion), it is better and scientifically more accurate to use an evidence-based reasoning. So, for example, writing:

We show through a thorough analysis that our model result is consistent with observations

is more appropriate than:

We believe our model result to be true.

Another way of supporting arguments and statements is by citing specific facts and providing references wherever applicable.

Note, however, that being objective should not prevent you from communicating to your readers what you *think* of the results you present. In fact, as we shall see in Section 9.3.1, you are *expected* to share your views with your readers so that they can perceive your conclusions in a coherent way. Making sure that your own writing adheres to the many features of scientific writing is one of your main tasks during *editing*. This step is where you need to spend enough time to check that your language is unambiguous, clear, precise, and substantiated. Often this requires not only that we choose words carefully but that we get rid of unnecessary words or phrases that do not serve any specific purpose in our prose.

5.3 PEARLS OF WISDOM: ADVICE FOR A BETTER STYLE

If you read any of the classic manuals on good English writing, such as *The Elements of Style* by Strunk and White [22], or *On Writing Well* by William Zinsser [23], you will notice that much of the advice provided is as valuable today as when it was first published.

Clarity, simplicity, and coherence represent key features of good academic writing. In general, these can be achieved by a few targeted actions that can be summarised as follows:

- use *verbs* not nouns;
- use the *active voice* whenever possible;
- place subject and verb *close together*;
- use *short words* instead of long ones;
- *de-clutter* your text;
- avoid *negatives*.

In the following sections we will see what exactly some of these actions imply and how simple edits can make your text flow and read better.

5.3.1 Verbs in action

Verbs represent the most important element of any sentence. Technically, a sentence cannot be defined as such unless it contains a *verb* that describes the action and provides meaning to the sentence. Quoting from Helen Sword's book, *The Writer's Diet* [19]: *"Verbs power our sentences as muscles propel our bodies"*. She further classes verbs as either *weak* such as those static, un-descriptive ones that convey little sense of action or direction (*to do, to be, to have, to make*, are all good examples of weak verbs), and *strong* such as those dynamic ones that evoke imagination and propel a sentence forward. Using static verbs somehow deprives the verb from its key function and makes our sentences dull and uninteresting. A beautiful example is again provided in Helen's book [19]:

> What is interesting about viruses is that their genetic stock is very meagre.

Clearly, there is nothing grammatically wrong with this sentence. However, it contains three instances of the verb *to be* (is) which is rather flat and does not convey any specific sense of action. Contrast with the following [19]:

> Viruses *originate* from a surprisingly meagre genetic stock.

Here, the meaning has been fully preserved, but replacing the verb *to be* with a more powerful **originate** leads to a major improvement in style. Similarly:

The announcement on the outcome of these applications will be made in July

becomes more elegantly expressed (and shorter) as:

> The outcome of these applications will be announced in July.

Table 5.2 Examples of strong verbs often found in academic writing.

agree	concur	disagree	explore	outline	review
analyse	contend	discuss	illustrate	persuade	revise
assert	contradict	dismiss	inspect	propose	suggest
assess	define	evaluate	investigate	purport	summarise
claim	demonstrate	examine	object	refute	support
clarify	describe	explain	observe	report	validate

It is often simple changes like these that can significantly improve your writing style. So, when editing your text, pay attention to the verbs you use. Identify any weak verbs, especially those that keep propping up all too often, and try to replace them with stronger ones that better convey the intended meaning. Of course, this may require a full reworking of your sentences and the way in which they link to preceding and following ones. Some useful strong verbs often used in scientific academic writing are listed in Table 5.2.

5.3.2 Verbs in disguise

Many of the terms used in scientific writing are highly technical and often abstract[2]. Some are so specific to a given subject that it would be impossible to replace them without ending up in long-winded ways of trying to express similar concepts. Provided a clear definition is offered for such terms, there is often no alternative but to use them whenever needed as the most accurate word in a given context.

Other abstract nouns, however, are used extensively whereas ideally they should be replaced with the verb they disguise. This is the case, for example, of many nouns ending in -*(t)ion*, -*ment*, -*ance*, -*ness*, -*ism*, -*ty*, such as *recommendation, argument, measurement*, which actually come from verbs such as *to recommend, to argue, to measure*. Overusing these abstract words makes for a heavy reading and does not aid understanding. Consider the following example:

---

[2] For further details on different types of nouns, see Chapter 10.

A two-stage detector was specifically built for light charged-particle detection at low energies and for the fulfilment of specific cross-section measurement requirements.

Here, the nouns **detection** and **fulfilment** may seem rather innocuous but they are abstract (and so are **measurement** and **requirements**) and therefore more difficult to understand than the verbs they hide. Far better is to rephrase the sentence by using the corresponding verbs instead, so as to obtain a sentence that is more elegant and easier to read, as in:

A two-stage detection array was specifically built *to detect* low-energy light charged particles and *to fulfil* the requirements of specific cross-section measurements.

A useful list of abstract nouns that are in fact verbs in disguise is given in Table 5.3. Note how these abstract nouns are often accompanied by static verbs such as *to make*. To replace the abstract noun with its corresponding verb, you may have to restructure your sentences, but both the style and clarity of your text will greatly improve as a result.

5.3.3 Active or passive?

Most scientific writing is about hard facts and objective evidence. This means that the focus of our writing is not so much on *who* did what, but rather on *what* was done or obtained. That is why the passive form[3] is widely used in scientific literature, especially in methods sections where procedures are described without specifying who carried them out.

Occasionally, passive constructions can also be used to maintain the same or similar subjects throughout a complex sentence, as in:

X-ray bursts have been explained (passive voice) as thermonuclear runaway on the surface of neutron stars accreting material from a less evolved companion and represent (active voice) one of the most common explosive events in the universe.

Here, the passive construction used in the first sentence allows the writer to keep the same subject, **X-ray bursts**, for the second sentence as well, but this time using an active construction.

Alternatively, passive constructions can be used to give emphasis to some words by placing them in a strategic position within the sentence. In general, however, the passive voice is longer, more static and also more difficult to understand and ideally you should use it sparingly.

By contrast, the active voice is shorter, more dynamic and also more trans-

---

[3]See Section 10.1.8 for more details on the difference between active and passive voices.

Table 5.3 Abstract nouns often used in scientific writing. For better style, replace the nouns with the verbs they disguise whenever possible.

(Verb +)	Abstract Noun	Verb
(make a)	recommendation	recommend
(formulate an)	argument	argue
(make a)	statement	state
(provide a)	description	describe
(find a)	solution	solve
(find an)	agreement	agree
(offer a)	suggestion	suggest
(make a)	measurement	measure
(create a)	contradiction	contradict
(show a)	disagreement	disagree
(provide a)	clarification	clarify
(make an)	objection	object
(make an)	analysis	analyse
(lead to a)	discussion	discuss
(make a)	summary	summarise
(provide an)	outline	outline
(provide an)	explanation	explain
(make an)	evaluation	evaluate
(make a)	revision	revise or review

parent as it clearly identifies the source of an action. So, unless you have good reasons for using the passive voice, as for example when what was done is more important than who did it, use the active voice instead. Whenever in doubt, opt for whichever construction conveys the clearest meaning. Chances are it will also make for a more elegant style.

5.3.4 *I* or *we*? Personal pronouns in scientific writing

A rather common misconception, and perhaps one of the reasons people advocate for the passive construction in scientific writing, is to assume that using the active voice automatically forces you to use personal pronouns such as *I, we, us,* .... As demonstrated by the previous example, this is not necessarily the case.

Yet, when it comes to the use of personal pronouns in scientific writing, opinions differ. And while most people would agree that the use of the personal singular pronoun *I* should normally be avoided, this may well depend on your discipline or even on the language in which you are writing. For example, when I was in Germany I noticed that the use of *I* in doctoral theses was the norm rather than the exception. In fact I would argue that the use of *I* is sometimes *needed* in a PhD thesis as a way to identify the specific contribution of the student as opposed to that of someone else (for example another student or researcher within the student's group or collaboration).

Different is the case in scientific research papers. Here, the use of *I* is almost universally avoided, although *we* is sometimes allowed as a replacement, even when a single author is intended.

Ultimately, the choice of whether to use personal pronouns or not rests with the author(s). Occasionally, a good solution might be an impersonal construction, with *it* as a subject, as in **It was decided**.... Indeed, science students are often advised to use the passive construction as a way to make their prose impersonal. Just make sure you do not overuse it.

5.4 DE-CLUTTER YOUR TEXT

Some journals place strict limits on words or page numbers. Occasionally, even referee reports come back to the authors with the instruction of cutting the text by 20% or 30%. Similarly, forms used for research grant applications often provide a strictly limited space for the applicants to make their points. Learning how to cut down your text without losing content is thus an art to be mastered.

When I was an undergraduate student, realising that learning English would be important for my career, I decided to take private tuition from a native speaker. As it turned out, my teacher, Ms. Buday, was one of the best I have had throughout my education and most of what I know about English I owe it to her. One of hardest, yet most useful, exercises she used to give us consisted in writing a one-paragraph *precis*[4] for every passage we read in the class. As the number of words we could use for our precis was strictly limited to 50, we had to learn how to extract key information from the passage, typically 400-450 words long, and to leave out any unnecessary detail. In hindsight, I believe this type of training has given me the ability to summarise texts quite effectively, while also teaching me what type of information is of secondary importance.

Admittedly, when it comes to our own writing, we feel very protective of what we have drafted and the idea of cutting *any* text we have spent so long to compose becomes a rather daunting prospect. Incidentally, that is why sometimes editing is best done by someone who is not emotionally attached to the text.

In the absence of a skilful and trusted colleague who does the job for you, you will have to be ruthless about what you have written. The best way I have found to edit my own drafts is to cut unnecessary text and paste it into a separate file so that I can re-use the material later on, if needed. The truth is, I never do.

---

[4] A precis is a concise summary of essential points, statements or facts from a longer text.

When editing your drafts, consider whether all you have written is strictly necessary to convey your message. You may be surprised by how much text you can cut without losing content. As we have seen, using the active voice is one way of significantly reducing the overall length of your text. Others include eliminating waste words, avoiding repetitions, and using affirmative statements, as discussed below.

5.4.1 Awkward phrases and waste words

An effective way of polishing your text during editing is by eliminating superfluous material that does not add new content or information. This requires identifying and eliminating empty words or phrases, such as *there is, there are, it should be stressed that, as can be seen* that do not actually mean much, and going instead straight to the point. So, for example, rather than saying

> There seems to be some evidence to suggest...,

opt for a simpler and crispier:

> Some evidence suggests....

Of course, you may argue that the level of caution in the original sentence is not retained in the revised version. However, when recognising that the verb *suggest* already contains an element of doubt, you may accept that the intended meaning is preserved.

Similarly, long-winded ways of saying something can often be expressed more clearly as single words, as in the examples below:

an adequate amount of	enough
at the present time	now
by means of	by
due to the fact that	because
in case of	if
in order to	to
in the absence of	without
in the course of	during
in the vicinity of	near
located in; located at	in; at
with a view to	to

To keep your style trim and fit try to avoid overusing awkward phrases and replace them with simpler expressions wherever possible.

5.4.2 Transition words

As we have seen in Section 4.6, transition words, such as *however, therefore, nevertheless*, or phrases as *in conclusion, to summarise, on the other hand*, are often used to connect sentences and paragraphs together. They are helpful to structure our writing and act like road signs to guide our reader through complex developments of ideas. However(!), overusing transition words can obfuscate the meaning and bury your key message in a muddle of unnecessary complications.

By contrast, words such as *obviously, undoubtedly, clearly, unique*, if used sparingly, can provide emphasis, convey confidence and instil a sense of trust in your reader – a crucial factor in persuading our referees, especially in grant proposal applications.

5.4.3 Redundant information

As I was editing an earlier draft of this chapter, I noticed that one of the tables' titles read:

List of some abstract nouns often used in scientific academic writing that are actually verbs in disguise and should ideally be replaced by their corresponding verb. (26 words)

This title was simply too long and wordy. So I reasoned that I could have easily got rid of **List of some** as it was clear that I was presenting a list and it was obvious that the list was by no means exhaustive. I then acknowledged that I could drop **academic** without losing content, and I could express more concisely the next two clauses **that are actually verbs in disguise and should ideally be replaced by their corresponding verbs**. The revised version now reads:

Abstract nouns often used in scientific writing. For better style, replace the nouns with the verbs they disguise whenever possible. (20 words)

Do not shy away from cutting down any redundancy, needless words, or repetitions. Not only will your sentences be shorter, they will also read better.

5.4.4 Negative statements

Negative statements in English are expressed in one of three possible ways: by adding *not* to the main verb; by using negative nouns, adjectives or adverbs (such as *never, nothing, none, no-one*); or by adding a negative prefix such as *un-, non-, in-, dis-*, to the affirmative word (for example: *un*-suitable, *non*-existent, *in*-accurate, *dis*-connected).

When two negatives are used together, they somehow cancel each other and become the equivalent of an affirmative. So, for example, the phrase **it is not unusual** actually means **it is usual**. True double negatives, however, such as **I did not know nothing**, are plainly wrong in standard English, though they may occasionally appear in songs or in colloquial expressions. This is not the case in other languages, though, where in fact double negatives are perfectly fine and still have a negative meaning (this is called *negative concord*). So, for example, the Italian *"Non ho fatto niente"* (literally, **I did not do nothing** – a phrase often used by my 7-year-old son) actually means **I did not do anything!**

This reminds me of a funny post I once read on Facebook explaining how in certain languages (like English) the use of two negatives actually implies something affirmative, while in others (for example, Italian) a double negative remains a negative. By contrast – the post continued – there is no language in the world where two affirmatives make up a negative. And then someone commented: *"Yeah, sure!"*.

Apart from the funny joke, the main problem with negatives in academic writing is that they can generate confusion and ambiguity, especially for non-native speakers of English who may be baffled by the meaning of negative sentences. Consider, for example, the following sentence:

Although these results are not in good agreement with simulated values, the discrepancy was regarded as not significant as mostly due to the many approximations made regarding the position of the alpha source. (33 words)

Here, the presence of two negatives (**not in good agreement** and **not significant**) together with the concessive clause introduced by **although** make the sentence cumbersome to read and understand. Better is to rephrase it in the affirmative, as:

Because of the many approximations made about the alpha source position, we regard the results to be in fair agreement with simulations. (22 words)

Finally, like passive voice constructions, negative statements require a little extra time to be processed by our brains and thus slow down understanding. Ideally, you should use negatives sparingly or for added emphasis, as in the famous statement by James Watson and Francis Crick at the end of their paper proposing the double helical structure of DNA [*Nature*, vol. 171 (1953) 737-738]:

It has not escaped our notice that the specific pairing we have postulated immediately suggests a possible copying mechanism for the genetic material.

Here, the use of **It has not escaped our notice** emphasises the authors' awareness about the momentous implications of their discovery in a way that a mere **We notice** would not have.

Chapter 5: The Editing Step
In a nutshell...

- Editing is where writing meets elegance. Be prepared to spend enough time to make your writing pleasant to read.

- Pay constant attention to the style used in your discipline and familiarise youself with its key features.

- Assess the most pressing stylistic issues of your writing (for example by taking the *Writer's Diet Test*) and edit your text accordingly.

- Favour strong and dynamic verbs to static and boring ones. Limit the use of abstract nouns that could be better replaced by verbs.

- Use the active voice whenever possible.

- De-clutter your text. Eliminate any empty words and long-winded expressions that do not add content or information.

- Avoid negatives unless necessary or for added empahsis.

EXERCISES

5.1 **Replace weak verbs with strong ones** [20 minutes]. Going through a section at a time, underline all verbs, whether active or passive. Identify instances where you use weak verbs such as *do, make, be, have* that are used in their own right (rather than as auxiliaries). Hence try to see if you can replace most of them with a strong verb, as one of those listed in Table 5.2.

5.2 **Verb in disguise** [20 minutes]. Look for all abstract nouns ending in *-ance, -ment, -ness, -(t)ion, -ty* and see if you can replace them with the verb they disguise. Of course you will need to re-arrange the structure of your sentences.

5.3 **Change passive into active** [30 minutes]. Identify all the passive constructions in your section. Hence try to assess whether the passive is justified (this is often the case in methods sections). If not, try replacing the passive with the active voice.

5.4 **Cut the clutter** [20 minutes]. Following the advice in Section 5.4, identify any empty words or redundant information that you can cut without losing content. Try to reduce the overall length of your section by about 20%.

5.5 **Avoid over-using negatives**. Locate negative statements in your text and see whether you can convert them into affirmative ones.

FURTHER READING

Atkinson, I. (2011) *Copy. Righter.*, LID Publishing Ltd.

The Academic Phrasebank
  http://www.phrasebank.manchester.ac.uk

HELPS (Higher Education Language and Presentation Support)
  https://www.uts.edu.au/current-students/support/helps/about-helps
  ssu.uts.edu.au/helps

RMIT - Learning lab: Writing skills
  https://emedia.rmit.edu.au/learninglab/content/writing-skills
  https://emedia.rmit.edu.au/learninglab/sites/default/files/academic_style.pd

University of New South Wales
  Online academic skills resources
  https://student.unsw.edu.au/writing

Monash University - Language and learning online
  http://www.monash.edu.au/lls/llonline/writing/science/index.xml

University of Wollongong - UniLearning
  http://unilearning.uow.edu.au/main.html

On the discovery of the structure of DNA
  http://www.exploratorium.edu/origins/coldspring/ideas/printit.html

CHAPTER 6

The Proofreading Step

CONTENTS

6.1	When details matter	84
6.2	Common grammar mistakes	84
	6.2.1 Homophones	84
	6.2.2 Subject-verb concordance	85
	6.2.3 Singular or plural?	85
	6.2.4 Dangling participle	86
	6.2.5 Ambiguous referencing	87
6.3	Punctuation marks often misused	88
	6.3.1 Comma	88
	6.3.2 Semi-colon	90
	6.3.3 Colon	90
	6.3.4 Hyphen	91
	6.3.5 Apostrophe	91
6.4	Spelling checks	92
6.5	Citations and bibliographies	93
	6.5.1 What to cite, where, and how	94
	6.5.2 Reference formats	94
6.6	Proofreading checklist	96

PROOFREADING is a crucial step of the writing process and it pays off to do it well. Sadly, it is also often overlooked by most students.

Some people tend to identify proofreading with spellchecking but this is only one aspect of it. The key purpose of proofreading your work is to detect and correct not just typographical errors but also any mistakes in grammar, punctuation, and references. It also implies checking that your text is formatted consistently and adheres to specific editorial conventions, as prescribed by different journals and universities guidelines.

By its nature, proofreading is normally a very meticulous task that requires great attention to details. Ideally, it is best done by someone who is not familiar with the text and so you may wish to enlist the help of a fellow student or colleague (just make sure you reciprocate!). But if you decide to proofread your own work, you should ideally do so some time after having finished writing so that you can spot mistakes more easily.

Also, make sure you have fully completed all previous stages in the writing process: there is no point in proofreading your work if you are still creating content or polishing your style.

6.1   WHEN DETAILS MATTER

Some time ago, a colleague of mine told me he had received a PhD thesis to examine. As he opened the thesis to have a quick look at it, he was baffled to see that all references in the bibliography at the end of the thesis had the same journal name, volume and year. Only the author names were different! Clearly, the student in question must have run out of time as s/he approached the submission deadline and made a quick cut and paste without changing the bibliographical records for each entry. As my colleague readily admitted, he quickly formed an impression of the student in question as not being very professional.

Unfortunately, this is not an isolated example. I have often read theses or reports littered with grammar mistakes or spelling typos. Minor as these issues may seem at first, they often portray a very unfavourable impression of the author as someone careless who did not take time to do things properly. Of course, we all make mistakes and it is almost unavoidable to overlook a typo or two, especially in lengthy documents such as a PhD thesis or a long paper. Yet, when such issues recur frequently in a text, the reader is invariably left wondering about the accuracy and care that the student has put in his research and whether his results can really be trusted. This is obviously not a position you want to be in. You may be surprised to realise how many readers will unconsciously judge how professional you are based on the quality of your production!

So, after these words of warning, let us have a look at exactly what to do when proofreading your work.

6.2   COMMON GRAMMAR MISTAKES

Grammar mistakes are among the most annoying problems of an academic text. I am aware that for most people English is not their native language, and we all make mistakes from time to time. This is unavoidable. However, a text full of grammar mistakes becomes difficult to understand, it will not get your work published and, worse – as we have seen – it can kill your credibility.

Yet, many mistakes are really easy to spot and can be rectified upon proofreading. Some of the most common ones are listed in the following sections.

6.2.1 Homophones

Homophones are words or groups of words that share the same pronunciation but have different meaning and spelling. Perhaps, the most common mistake I typically come across (whether in academic writing or not) consists in the wrong use of *it's* and *its*. Luckily, this is also one of the easiest to correct. Simply remember that *it's* indicates a contraction and stands for *it is* or *it has*, while *its* indicates possession and stands for *of it*. Do not assume that you can use them interchangeably just because some people do. You cannot!

To a similar category belong the mis-use of *you're* and *your*; and *they're*, *their* or *there*. Again, the apostrophe in such cases indicates contraction and not possession. So, for example *they're* means *they are* and should not be confused with *their* meaning *of them* (while *there* is the opposite of *here*). We will explore other uses of the apostrophe in the following section.

Unlike homophones, other words share similar meanings but have different grammatical functions. Examples include: *who* and *whom*; *less* and *fewer*; *due to* and *owing to*; *that* and *which*. These are discussed in more detail in Chapter 10.

6.2.2 Subject-verb concordance

Verb-subject concordance can be easily broken when using phrases such as:

>A large number of plots *were* analysed

From a grammatical point of view, this sentence is not correct because the word *number* is singular and as such requires a singular verb, namely:

>A large number of plots *was* analysed.

I am aware that some people may not like "the sound" of it. If so, my advice would be to avoid any controversies by rephrasing wherever possible as, for example:

>Several plots were analysed.

6.2.3 Singular or plural?

In a similar fashion, many words commonly found in scientific language are often used with a singular verb whereas in fact they ought to have a plural one. This is the case of many words from Latin and Greek origin, perhaps with the most common, and indeed controversial, being the use of *data* with a singular verb. *Data* is the Latin plural for *datum* (though datum is seldom used in its own right anymore). So, the correct form is:

>Data *were* collected...

and not:

>Data *was* collected...

Table 6.1 Singular and plural forms of commonly used words of Latin and Greek origin. Make sure they concord properly with the verb used.

Singular	Plural	Singular	Plural
analysis	analyses	hypothesis	hypotheses
addendum	addenda	locus	loci
appendix	appendices	matrix	matrices
axis	axes	medium	media
bacterium	bacteria	nucleus	nuclei
criterion	criteria	phenomenon	phenomena
datum	data	radius	radii
erratum	errata	spectrum	spectra
formula	formulae	thesis	theses

Admittedly, the singular usage has become so widespread that it is accepted by most. However, I would advice to use the locution *data set* (or *data point*, depending on context) if you really want to use a singular verb. Table 6.1 lists words commonly used in scientific context for which the plural form is often incorrectly used with a singular verb or vice versa.

6.2.4 Dangling participle

Participles are forms of a verb ending in *-ing* (present participle) or *-ed* (past participle) that can be used as adjectives to qualify a noun (for example: a bark-*ing* dog or a surpris-*ed* man). They can also be combined with prepositions and other nouns to form phrases which can be used to modify or describe the subject of the sentence. For example, in the sentence:

> Facing a hard deadline, I decided to work all night.

The participial phrase (**Facing a hard deadline**) clearly refers to the subject (**I**) of the main clause (**I decided to work all night**). However, if the subject referred to by a participial phrase is either not obvious or is omitted from the text, one is left with a *dangling participle*, as in:

Using a micrometer, the film was 3 μm thick.

Here the subject of the sentence is **the film**, but clearly a film cannot do the action of **using a micrometer**. As a result, the participial phrase is literally left dangling without referring to anything obvious. The way to fix such problems is by rephrasing the sentence so that either: a) the proper subject appears explicitly in the text; or b) the participle no longer refers to the subject of the sentence and therefore is not left dangling. In the previous example, the subject of *using a micrometer* is obviously a person (you, a student, a researcher), so the sentence could be rephrased either as:

Using a micrometer, we measured the thickness of the film to be 3 μm.

or:

The film was 3 μm, as measured using a micrometer.

6.2.5 Ambiguous referencing

Personal pronouns (*it, they, he, she, him, her, we, us,*...) and demonstrative articles or pronouns (*this, that, the, these, those,*...) are extremely helpful to avoid repeating the same noun or word, but if not used properly they too can give rise to ambiguity. This occurs when the pronouns do not concord (either grammatically or in meaning) with the noun(s) they are supposed to replace or when it is unclear *which* noun(s) they refer to (see also Section 5.3.4).

As your reader will not have the benefit of asking for clarifications, it is always critical to check that pronouns or demonstrative articles refer to the intended noun(s) and that the meaning is unambiguous.

At times, ambiguity arises from improper grammatical structure as illustrated by the following examples:

Each pulse was recorded for a total of 5 μs around the signal peak consisting of 1000 sampling points.

Here, it is unclear whether **consisting** refers to **the signal peak** or to **each pulse**, so the sentence should be rephrased as:

Each pulse was recorded for a total of 5 μs around the signal peak, which consisted of 1000 sampling points.

or, depending on the intended meaning:

Each pulse, consisting of 1000 sampling points, was recorded for a total of 5 μs around the signal peak.

Another example:

The target holder was water cooled in order to be able to control the temperature to minimise target degradation.

Here, the ambiguity comes from a discordance between intended meaning and grammatical function. That is, **in order to be able to control the temperature** grammatically refers to **The target holder**, but by meaning it clearly refers to the experimenters because the act of controlling the temperature cannot be performed by the target holder. By rephrasing the sentence as:

The target holder was water cooled to minimise target degradation effects

both ambiguity and poor concordance are resolved.

6.3 PUNCTUATION MARKS OFTEN MISUSED

Punctuation marks fulfil in writing the same function that pauses and intonation fulfil in speaking. As such, punctuation is essential to writing. Most common punctuation marks are easy to use: full stops indicate the end of a sentence; question marks are used to signal a question; exclamation marks to add emphasis (though this is rarely used in academic writing); capital letters for proper nouns, nationalities, languages, days of the week, months, public holidays, geographical locations and so on. Other punctuation marks, however, are often used improperly, most notably the *comma* (,); the *semicolon* (;); the *colon* (:); the *hyphen* (-); and the *apostrophe* ('). Their function and correct usage is discussed below.

6.3.1 Comma

Commas signal a brief pause and are used to separate items in a list or more generally to avoid possible ambiguities.

DO USE a comma in the following situations:

- **To separate items in a list.**
  Note that the final item in the list is often preceded by the conjunction *and* and a comma is normally not necessary. However, occasionally a comma is required before the *and* to avoid ambiguity. For example:

    I love my parents, Whitney Houston and Michael Jackson

  could be read to imply that Whitney Houston and Michael Jackson are your parents. Unless this is the case, use instead:

    I love my parents, Whitney Houston, and Michael Jackson

This use of a comma is referred to as *Oxford comma*.

- **After certain adverbs** (*however, certainly, nevertheless, in addition*).
  Remember though, that if you use *however* at the beginning of a sentence without a comma, it means "in whatever way", "to whatever extent", "no matter how". For example:

  However hard he tried, he never succeeded,

  meaning:

  No matter how hard he tried, he never succeeded.

- **To mark non-restrictive clauses**[1].
  Be aware that whether a comma is used or not can change the intended meaning. Compare for example the two following sentences:

  All students, who attended most lectures, passed the exam

  with

  All students who attended most lectures passed the exam.

  In the first case, all students passed the exam, while in the second one only those who attended most lectures did. So, depending on exactly what you want to say, make sure you use commas appropriately. Also, note that in the first example the clause in between commas is *incidental*, i.e., it merely provides additional information, and as such it can be removed without altering the meaning of the main clause.

- **To separate the main clause from a subordinate**[2] **one.**
  This is especially relevant if the subordinate clause precedes the main one. The use of the comma here is to signal to the reader where the subordinate clause ends, which can be especially helpful for long clauses. For example:

  Before being anodised in ^{18}O-enriched water (subordinate clause), the tantalum disks were etched in a bath of 20% citric acid solution (main clause).

- **To avoid confusion or ambiguity.**

---

[1] A non-restrictive, or *incidental*, clause is one that adds extra or non-essential information to a sentence. As the clause can be safely omitted without altering the meaning of the remaining sentence, it is normally enclosed between commas.

[2] A subordinate, or *dependent*, clause is one that merely provides additional information to the main clause, but cannot stand alone.

DO NOT USE a comma in the following situations:

- **After a conjunction** (*and, but, or*).
  The use of a comma is allowed, however, if the conjunction is followed by an incidental clause.

- **Between subject and verb, or verb and object.**
  In very specific circumstances, the use of a comma between subject and verb can be tolerated, especially with incidental clauses, for example:

  > Our result, on the contrary, was more accurate but less precise.

- **To separate a subordinate clause that follows a main one.**
  When the main clause precedes the subordinate one, there is no need to place a comma to separate the two as it is normally clear where the main clause ends. Using the same example above, and inverting the order:

The tantalum disks were etched in a bath of 20% citric acid solution (main clause) before being anodised in ^{18}O-enriched water (subordinate clause).

6.3.2 Semi-colon

Use a semi-colon to:

- **Separate items in a list.**
  This is especially important if the items already contain commas.

- **To separate two closely related independent clauses.**
  The wrong use of a comma in such instances is referred to as *comma splice* and should be avoided. For useful advice on various ways to remedy a comma splice you can refer, for example, to a brief blog post at: https://en.oxforddictionaries.com/grammar/the-comma-splice.

6.3.3 Colon

Use a colon to introduce or announce:

- **Examples**

- **Explanations**

- **Lists**

- **Direct quotations.**

Some people often use a semi-colon instead, but that usage of a semi-colon is not correct.

6.3.4 Hyphen

Use a hyphen in the following instances:

- **To mark a relationship between two words and avoid potential ambiguity.**
  For example, use **old time-machine** or **old-time machine** depending on whether you refer to a time-machine that is no longer new or to a machine from old times, respectively.

- **To avoid unusual vowel combinations.**
  For example, **de-ice** or **re-examine**

- **To link together nouns used as adjectives.**
  For example, use This is a gamma-ray detector (with hyphen), but This is a gamma ray (without hyphen).

6.3.5 Apostrophe

The apostrophe is perhaps one of the most misused punctuation marks and it is important to understand its different functions to be able to use it correctly. As shown in Table 6.2, the apostrophe is used for one of three purposes:

- **Indicate omission (or contraction)**
- **Indicate possession**
- **Indicate abbreviations.**

Note that in academic writing you should avoid to use apostrophes for contractions and to spell out full words instead (for example, use **did not** instead of **didn't**). Doing so will also help you decide exactly what you want to say in situations when in doubt about the use of the apostrophe. Do you mean it is or of it? In the latter case, the correct form is **its** (possessive adjective or pronoun). Also be aware that, in some cases, the meaning changes depending on whether you use an apostrophe or not. For example: **1970's** means "of the year 1970", while **1970s** means "the seventies".

Do not use an apostrophe in plurals, acronyms, or with personal possessive adjectives and pronouns. So, for example, write **CDs** to indicate more than one compact disks, but write **CD's**, as in the **CD's cover**, to indicate possession (the cover of the CD). Similarly, never use **your's** to indicate possession. If possession is intended, the correct form is **yours** as in: I found my book but not yours. Similarly, use **theirs**, without apostrophe, to mean the one belonging to them.

Finally remember that **who's** and **whose** have the same pronunciation but they mean **who is** and **of whom**, respectively, and must not be confused nor used as if they were interchangeable: they are not.

Table 6.2 Apostrophe's functions.

Contraction	Possession	Abbreviation
can't = can not	Peter's = of Peter	gov't = government
I'd = I would, I had	Charles' = of Charles	cont'd = continued
it's = it is, it has	children's = of the children my pet's = of my pet my pets' = of my pets	int'l = international
who's = who is	butcher's (shop) dentist's (practice)	'less = unless 'til = until
you're = you are	1970's = of the year 1970	'70s = in the seventies
	five months' experience	

Talking about apostrophes, I remember a funny story that happened during one of my writing workshops. I was showing my students a picture of a poster I had seen at Edinburgh airport advertising free food for kids. The poster read: **Kid's eat free**. When I asked my students to point out what was wrong with that picture, someone from the audience shouted: *"They should pay!"*. I put my hands on my head in despair!

6.4 SPELLING CHECKS

Correct spelling is almost as important as correct grammar. The following list provides useful advice to refer to when proofreading.

- **Use a spell checker**
  These days, almost every text editor comes with a built-in spell checker, even in our mobile phones! In most cases, the software is very effective,

but remember that no spell checker will be able to spot and correct for homophones (Section 6.2.1).

- **Search for common mistakes**
  Using electronic searching capabilities, focus on words that you might likely mistype, such as *form* and *from*, or any that can be easily confused. A search tool can also be useful to check that all instances of *it's* and *its* are properly spelled and mean what they are supposed to. Similarly, you can use a spellchecker to rule out any *dangling participle* by searching for all *-ing* forms and checking that they are used correctly.

- **Proofread from a printout**
  Once all electronic spell checking has been completed, it is advisable to continue proofreading on paper, ideally covering with a blank sheet all the lines below those you are reading. This technique will help you avoid reading ahead and thus possibly skipping any mistake.

- **Read backwards**
  A useful way to help you focus on the spelling of each word consists in reading sentences backwards. This will help you focus on spelling rather than being distracted by meaning.

- **Beware of autocorrection functions**
  Some spellcheckers also provide an autocorrection function, but this is not infallible, to say the least. Always check that the replaced text makes sense and means what you intended.

- **Consult a dictionary when in doubt**
  This is self-explanatory, really. Yet, it always surprises me to realise how few people use a dictionary. Remember that many words in English have similar spelling but very different meanings: for example, **illusion** and **allusion**; **principal** and **principle**. For a list of commonly confused words in scientific writing, see Table 10.2 in Chapter 10.

Finally, remember that British and American English adopt different spelling for certain words (for example, *centre* and *center* in British and American English, respectively). In general, the choice of one or the other is dictated by whether a given journal or publishing house is one based in the UK or in the US (or elsewhere). If no restrictions apply, feel free to choose the one you prefer, but be consistent throughout your manuscript.

6.5 CITATIONS AND BIBLIOGRAPHIES

A key feature of academic writing is the presence of citations, and for good reasons. Citations are necessary to show that you have consulted the work of others and are familiar with the ongoing debate in your field. They also add weight and credibility to your statements and provide the reader with an opportunity to check the accuracy of the information you present. Finally, they serve to acknowledge other people's work and to avoid plagiarism.

6.5.1 What to cite, where, and how

Inexperienced writers often express doubt as to what to cite, where, and how. As a rule of thumb, you should focus only on literature that you have actually read and cite only those works that are *necessary* and *sufficient* to provide your reader with a comprehensive overview of your research project and how it fits within the broader context of your discipline. Ideally, you should cite *primary literature* wherever possible, that is the original papers or sources where the information first appears. If works are very well established, too old or otherwise difficult to trace, citations to the *secondary literature* that refers to the original work is also a valid alternative, especially if this latter is the only literature that you have read.

When providing references in your text, make sure they are placed exactly where they belong. This is normally at the end of some general statements or considerations or immediately after a specific statement or result taken from a given source, as in the example below.

> Classical novae, a frequent phenomenon in our Galaxy, are explained as thermonuclear explosions on the surface of white dwarf stars accreting hydrogen-rich material from less evolved companions in binary star systems [1] and have been proposed as a key source of ^{13}C, ^{15}N, ^{17,18}O [2] and ^{18,19}F [3] isotopes in the Universe.

Finally, remember that not all citations are equally reliable. For example, internet sources are not appropriate academic references as they do not typically undergo the severe peer-review scrutiny reserved to published papers. As such, these should be used cautiously.

6.5.2 Reference formats

When it comes to formatting references, you need to be aware of any specific requirements of your university (if you are submitting a PhD thesis) or the journal to which your intend to submit your paper. Note, also, that different disciplines may adhere to different formatting conventions. Make sure you are familiar with good practice in your own field.

In scientific writing, the most widely used referencing formats are the *Vancouver system* and the *Harvard system*. In the Vancouver system, references are numbered consecutively using Arabic numerals enclosed in square brackets, for example "[3]". At the risk of stating the obvious, please make sure you start numbering from "[1]"! In the Bibliography, at the end of your thesis or paper, all references are reported progressively as they appear in the text. Note that for a thesis, the Bibliography should start on a new page.

In the Harvard system, references are cited in the text by the surname (without initials) of the first author (or the first two or three if the paper has two or three authors at most) followed by the year in brackets, for example: **Smith (2012)** or **Smith and Anderson (2014)**. For articles with more that three authors, it is customary to add *et al.* in italics after the first author name, as in **Smith *et al.* (2012)**. For different papers by the same author(s), citations are given with the different years of publication separated by a comma: **Smith and Anderson (2014, 2015)** or **Smith (2012a, 2012b)**. In the Bibliography, references will appear in alphabetic order and, for same author papers, in chronological order, regardless of the order in which citations appear in the main text.

Formatting rules for the items reported in the Bibliography depend on the type of source you are citing and may vary considerably across disciplines. The list below provides examples of typical formatting styles for most common sources used in an academic context. Always check if any different conventions or guidelines apply in your field.

- **Journal articles** are normally cited with author names first, followed by the journal name, the volume (typically, though not always, in boldface), the year in round brackets, and the article's first page or page range. For example:

J.K. Smith *et al.*, Physical Review Letters 117 (2016) 331.
J.K. Smith *et al.*, Physical Review Letters **117** (2016) 331-335.
J.K. Smith *et al.*, Physical Review Letters 117, 331 (2016).
Smith, J.K., *et al.*, Physical Review Letters 117, 331 (2016).

Occasionally, you may also find journal articles cited as:

Smith, J.K. (1997) Journal of Higher Education 14 (3) 118–123.

Here, the first number after the journal's name indicates the volume, the number in brackets the issue, and the last numbers the article's pages.

- **Books** are normally cited with the name of the author, the year of publication, the book title (in italics) and the publishing house:

Brown, G. (2000) *Effective counselling*, Wiley & Sons.

If the book has more than one edition, also state the edition you have consulted. If you refer to specific pages, sections or chapter(s) in the book, state those explicitly at the end of the citation, as in:

Brown, G. (2000) *Effective counselling*, Wiley & Sons, chap. 3; or
Brown, G. (2000) *Effective counselling*, Wiley & Sons, pp. 77-94.

- **Theses or dissertations** are cited by providing the author, the year, the title (in italics) and the university, as in:

Scott, D.A. (2012) *The $^{17}O(p,\gamma)^{18}F$ reaction in classical novae*, PhD Thesis, the University of Edinburgh.

- **Internet sources** should include the author name (if known), the full document title (if available), the full URL, the date of publication and, possibly, the date you accessed the information. This latter can simply be set to the date you perform the final check before submitting your paper or dissertation (just make sure the site is still available then). For example:

Micron Semiconductors Ltd., http://www.micron.co.uk (15 May 2012)

6.6 PROOFREADING CHECKLIST

Here is a useful checklist for you to go through when proofreading your text.

PROOFREADING CHECKLIST	
1. Text checked for typos using a spellchecker	[ ]
2. Text checked for common grammar mistakes	[ ]
3. Punctuation used correctly throughout	[ ]
4. All references formatted consistently and according to requirements	[ ]
5. Numbered references appear in the right sequence, starting from 1	[ ]
6. All figures and tables are recalled in the main text	[ ]
7. Captions and titles contain enough information for stand-alone figures and tables	[ ]
8. All labels in figures are clearly defined and units given where needed	[ ]
9. Different lines and symbols can be distinguished when printed in black and white	[ ]
10. All proper names are spelled correctly (e.g., in the Acknowledgments)	[ ]
11. Figure resolution is high enough for possible size reduction	[ ]

Chapter 6: The Proofreading Step
In a nutshell...

- Proofreading is a crucial part of the writing process. Do not overlook it and leave enough time to thoroughly check your final text.
- Familiarise yourself with reference formatting conventions in your discipline or check the authors' guidelines provided by journals.
- Pay special attention to any grammatical error, spelling or formatting mistake. If in doubt consult a dictionary or a grammar book. Non-native speakers of English may want to check the quality of their English with a native speaker.
- Use a spellchecker but do not rely on it alone. No spellchecker can pick up on words with different spellings, both of which correct (for example, *form* and *from*).

EXERCISES

6.1 **Backward reading**. Reading your text backwards, from the last line to the first, can help you to locate more easily any typos as you will be less concerned with the meaning of what you are reading. Of course, this will not allow you to pick up on grammatical issues such as subject-verb concordance.

6.2 **The English friend**. If you are not a native speaker of English, consider enlisting the help of a trusted native speaker friend who can check the language (including grammar) for you.

6.3 **Checklist**. Use the checklist at the end of this chapter to make sure you have not forgotten any important check. Add your own items to the list if you wish.

FURTHER READING

Truss, L. (2009) *Eats, Shoots & Leaves – The Zero Tolerance Approach to Punctuation* Fourth Estate, HarperCollins.

Holtom, D. and Fisher, E. (2006) *Enjoy Writing Your Science Thesis or Dissertation!* Imperial College Press.

Academic Writing Help Center – University of Ottawa
http://sass.uottawa.ca/en/writing/resources

Brians, P. (2003) *Common Errors in English Usage*, William, James & Co.
https://brians.wsu.edu/common-errors/

Common Grammar Mistakes
    https://www.copyblogger.com/grammar-goofs/
    https://blog.hubspot.com/marketing/common-grammar-mistakes-list
    https://marialuisaaliotta.wordpress.com/2012/04/28/its-its-isnt-it/

Apostrophe abuse (fun guaranteed!)
    http://www.apostropheabuse.com

CHAPTER 7

The Technical Stuff

CONTENTS

- 7.1 Titles 99
- 7.2 Table of Contents 101
- 7.3 Figures and Tables 102
  - 7.3.1 What goes in a figure 102
  - 7.3.2 Figure captions 103
  - 7.3.3 Examples of poorly prepared figures 104
  - 7.3.4 What goes in a table 106
  - 7.3.5 Table titles 107
- 7.4 Equations and symbols 108
- 7.5 Reporting experimental results 109
- 7.6 Appendices 109
- 7.7 Glossary and Lists of Acronyms 110
- 7.8 Acknowledgments 111

AN EXCELLENT compendium on all matters associated with the preparation of figures and tables, symbols and formulae, style issues and formatting conventions is presented in the American Institute of Physics *AIP Style Manual* [24] and can be downloaded online from various websites. The manual offers sound general advice for the preparation of clear, concise and well-organised manuscripts and even if it targets specifically researchers in physics and astrophysics, many of the guidelines presented are equally valid for scholars in other scientific disciplines.

For other discipline-specific conventions, you may want to refer to similar manuals in your own area (some links are provided in the Further Reading section of this chapter). Here, we present a brief overview of some key elements for you to keep in mind when preparing any supporting materials for your writing.

7.1 TITLES

Titles are among the most important elements of academic theses or research papers. Often, a title is the only basis upon which a potentially interested reader will decide within a split second whether (s)he wants to invest time and effort into reading any further. Ideally, a well-written title should indicate the *nature* and *purpose* of your research, should be brief and accurate, and should contain key words or concepts that other scholars would easily recognise. Fulfilling all of these requirements within a limited number of words is clearly challenging. So it is important that you spend some time playing around with several options before settling on a final title.

Some years ago I was working on a technical paper initially drafted by a colleague of mine. The working title was:

Realisation of isotopic oxygen targets by tantalum anodic oxidation.

At first glance, the title appeared to be a good one: it gave a hint to the fact that these targets were prepared by us; it mentioned that we were referring to oxygen targets; and it indicated the method by which these targets had been prepared (anodic oxidation). Upon reflection, however, I realised that the title lacked some key details. For example, it was not specific about the type of targets we were preparing; nor did it mention that we had also performed measurements to: a) determine their thickness and stoichiometry, and b) characterise their performance under ion beam bombardment. Finally, it did not say anything about the type of readership who might have been interested in our paper.

So, I decided to play around with various alternatives, by adding more information, making it slightly more specific, and re-arranging some of its elements. Eventually, as a compromise between length and accuracy, I settled for:

Preparation and characterisation of isotopically enriched Ta_2O_5 targets for nuclear astrophysics studies.

In this way, we were being specific about the purpose of the study and its experimental nature; we made it clear exactly what targets we were talking about by providing their chemical formula; and we addressed a specific audience for whom the paper might have been of relevance. Admittedly, I had to sacrifice the piece of information about the method but I decided that this was an additional detail that the interested audience (now clearly identified) would have discovered anyway as soon as they read the abstract.

Another example, was for the thesis of a PhD student of mine. The initial working title had been:

The $^{17}O(p,\gamma)^{18}F$ reaction in classical novae.

Again, despite being nice and short, the title was a bit too generic: yes, it did mention the specific reaction of interest and its relevance to a specific astrophysical scenario (**classical novae**), but it provided no clue as to the nature of the project (was it theoretical, computational, experimental?) nor to the novelty of the study. Eventually, we decided to go for:

First direct measurement of the $^{17}O(p,\gamma)^{18}F$ reaction cross section at Gamow energies for classical novae.

Admittedly, the title was much longer but it was also more accurate and specific: it made it clear the study was an experimental one (**measurement of... cross section**); it spelled out its importance by stating it was the **first direct measurement... at Gamow** energies (a technical term that people in the field of nuclear astrophysics would have immediately recognised); and it framed the study within the specific context of **classical novae**.

Crafting good titles is crucial to signal what kind of content your reader can expect to find in your PhD thesis or research paper. However, the choice of a good title is important not just for the main document, but also for its various components: chapters, sections and sub-sections. These, in turn, will help you to signpost the various aspects of your research in the Table of Contents (see below).

Clearly, the best time to decide on the titles of each section or document is after these have been completed, so make sure you take some time to revise any working titles as part of the proofreading stage.

7.2   TABLE OF CONTENTS

A carefully planned *Table of Contents* (ToC) is the best tool you can offer your readers to help them navigate through a dissertation, a thesis, or any extended report[1], as it provides an overview of its intrinsic structure at a glance. Once the reader recognises this structure, (s)he is in a better position to be able to follow the text as it unfolds.

In preparing a well-organised ToC, it helps if you think about it from your reader's perspective. For a ToC to be informative and helpful, chapters, sections and sub-sections must all have descriptive titles that are not limited to a generic *Introduction* or *Data Analysis and Results*. In other words, you should make it easy for your reader to know exactly what they can expect to find in that chapter, or section, or sub-section. So, for example, if you are writing about *novae* rather than opting for:

1.1 Introduction to stellar evolution

for the introductory chapter of your thesis, you could use instead:

1.1 Principles of stellar evolution in binary star systems

which is far more specific and accurate. An excerpt from a well-prepared ToC is shown in Figure 7.1.

---

[1]Tables of Contents are not normally used for scientific papers, unless – occasionally – for extensive reviews.

3	**Current Status of the $^{17}O(p,\gamma)^{18}F$ Reaction**	**32**
	3.1 Non-resonant component of the $^{17}O(p,\gamma)^{18}F$ reaction	32
	3.2 The $E_R = 183$ keV resonance	37
	3.3 The $^{17}O(p,\gamma)^{18}F$ reaction rate	38
	3.4 Motivation for present study	38
	3.5 Later measurements of the $^{17}O(p,\gamma)^{18}F$ reaction	40
4	**Experimental Approach**	**43**
	4.1 Stopping power and energy loss	43
	4.2 Yields and cross sections for charged-particle-induced reactions	46
	4.2.1 Non-resonant reactions	48
	4.2.2 Resonant reactions	50
	4.3 Gamma-ray peak shape for primary transitions	51
	4.4 True coincidence summing effects	55
5	**Experimental Setup**	**58**
	5.1 Cosmic background considerations	58
	5.2 The underground LUNA facility	60
	5.3 Experimental Apparatus	61
	5.4 Tantalum oxide targets preparation	64
	5.5 Activation measurements	70

Figure 7.1 Excerpt from the Table of Contents of a PhD thesis on the study of a nuclear reaction in Classical Novae. Note how descriptive titles provide a good indication of what the reader can expect to find in each section and sub-section.

7.3 FIGURES AND TABLES

Figures and tables form key ingredients of any scientific writing, as they can provide a large amount of information in a compact and efficient way. As the old saying goes, *a figure is worth a thousand words*. However, for figures and tables to be effective, some rules must be followed in their preparation. Perhaps, the most important feature of figures and tables, one which is often overlooked by novice writers, is that they are meant to *stand alone*. That is, readers should be able to understand what is in the figure (or table) without having to read the whole paper or chapter. In this way, they can decide whether the study is relevant to their own research before deciding to invest time in reading the entire document.

7.3.1 What goes in a figure

Most figures in scientific work consist in a two-dimensional graphical representation of the behaviour of some quantity (a dependent variable) on the y-axis with respect to another (the independent variable) on the x-axis. For figures to be unequivocally clear, they must fulfil the following requirements:

- Axes should be properly labeled, have units (if applicable) and these latter should be of an appropriate font size. In particular, you should ensure that all labels remain readable even if they are reduced in size during the production process for publication. Also, the scale limits of each axis should be carefully chosen as there is no point in leaving plenty of blank space in the figure that serves no purpose. This is especially important for research papers because journal space comes at a cost and it should not be wasted.

- The numerical values reported on each axis should be chosen with care: there is no point in quoting up to the second decimal digit, unless your measurement has the same sensitivity (see also Section 7.5). Similarly, for variables whose values span some orders of magnitude, a logarithmic scale may be more appropriate than a linear scale. Likewise, a scale in powers of 10 is easier to read and understand than the scientific exponential notation and it should be preferred.

- Define different symbols used to represent different data sets, either in the figure caption or in a legend. Where applicable, error bars should also be displayed and their origin (statistical, systematic, or both[2]) clearly stated.

- Avoid wasting ink by adding needless backgrounds to your figures. Widely used spreadsheets often add grey backgrounds by default, but this can (and should!) be easily overridden. Ideally, however, you should use more sophisticated data handling packages than spreadsheets to prepare high-quality figures that meet publishing standards.

- If using colours to distinguish among different data points or lines, always check that differences can still be appreciated if the figure is printed in black and white. Ideally, limit the use of colours as well as any 3D effects that do not add any useful information.

7.3.2 Figure captions

Despite being among the most important parts of a paper (or thesis), figure captions are often poorly written in that they do not contain enough information to make the figure stand alone. To avoid this pitfall, always include all relevant information to clearly state the meaning and purpose of the figure without requiring the reader to read the full paper/chapter.

The description provided in the caption need not be excessively long, but it should contain a number of essential components. These include:

---

[2] Broadly speaking, *statistical errors* are those resulting from fluctuations of measurements of the same quantity about the average; *systematic errors* are those resulting from the measuring devices used.

- a brief mention of what is shown in the axes (for example, **Experimentally measured yields [in counts/C] as a function of beam energy [in MeV]**) always stating what is on the y-axis *versus* what is on the x-axis, and not vice versa[3];

- some indication on how data were obtained;

- the meaning of symbols (appropriately labelled in case several data sets are shown); and

- the statistical significance (if applicable) of any numerical results.

For fairly complex figures, more words of explanation may be required. If the text gets too long, a phrase such as **see text for further details** can finally be added.

An excellent starting point for producing good quality figure captions is to identify well-written ones and simply pay attention to the type of information presented and how well the various components of the figure are described.

Finally remember that you are supposed to inform your readers about what you *think* of your results so that they can better follow your discussion and conclusions. As will be discussed in Section 9.3.1, this can be done in the main text by using appropriate language. Do not just assume that what is in the figure is clearly obvious: results do not speak for themselves and readers should not be expected to read your mind.

7.3.3 Examples of poorly prepared figures

Even though much of the advice given above seems obvious and unnecessary to state, it always amazes me to see how many students fail to produce good quality figures and captions. An example of a poorly prepared figure from a real case report on students' performance in *Advanced Levels*[4] is shown in Figure 7.2.

There are a number of issues with this figure, apart from the grey background. Despite stating **Decline in A Level standards**, the caption does not really explain what is in the figure, nor is it clear what exactly shows **a decline**. One may be led to assume that the decline is represented by the trend shown by both lines, but a closer inspection reveals that: a) the two lines are actually similar; and b) what is on the x-axis is not a continuous variable, but rather a sequence of distinct items, so that the figure should in fact display histograms, not lines.

Either way, the items on the x-axis are not defined, so the reader is left

---

[3] This is because the dependent variable (on the y-axis) is a function of the independent one (on the x-axis). The practice of stating "x vs y", often used by some native speakers of English, is not correct.

[4] Advanced Level, or A Level, is a secondary school leaving qualification in the United Kingdom (source: Wikipedia).

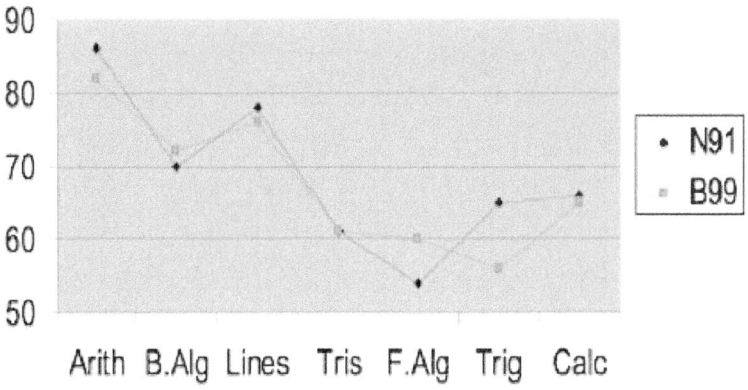

Figure 7.2 Example of a poorly prepared figure from a student's internal report showing a number of issues as discussed in the text.

guessing as to what those items represent: possibly something to do with arithmetic, trigonometry, or calculus?

Going back to the issue of the decline in standards, one might wonder whether this refers instead to a decline in performance, possibly across the years 1991 – 1999. This of course assumes that the two symbols N91 and B99 refer to two different years, but again neither of those symbols is defined in the legend or in the caption. Even in this scenario, however, it is difficult to see any decline in standards given that both lines are rather similar!

Maybe the decline is across subjects of increasing complexity, presumably from arithmetics to calculus, but this too remains guesswork in the absence of further indications from the writer.

Finally, the title above the figure (which in fact should not be there in the first place, because figures do not have titles) does not help either and it is not clear what the **N91 and B99 Diagnostic Test Results** are. Also, it remains unclear whether the numbers on the y-axis indicate percentage marks or fractions of population samples.

Another real-case example, this time of an apparently good figure and caption, is shown in Figure 7.3. Even though less severe than in the previous example, this figure too has its own issues. Specifically, the legend does not

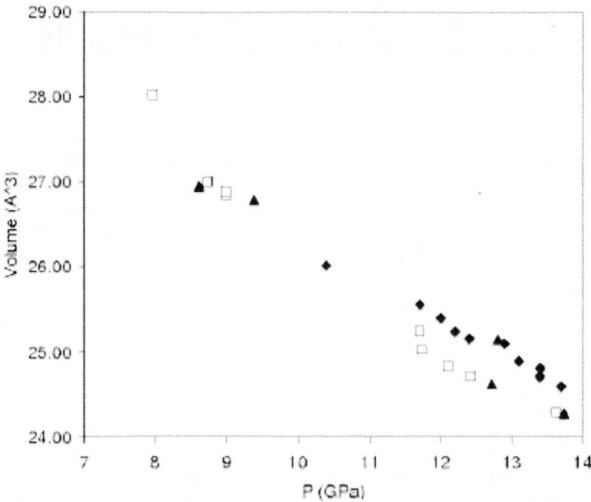

Figure 1: Plot showing variation of atomic volume vs. pressure for the stability range of Sm. Error bars corresponding to errors in volume and pressure (estimated) are omitted, as they are obscured by the symbols used to plot them the data.

Figure 7.3 Example of a poorly prepared figure from a student's internal report showing a number of issues as discussed in the text.

contain enough information to describe what is in the figure; the numbers on the y-axis are given with two decimal figures but for no obvious reason and their units are neither defined nor clear; the three different data symbols (triangles, open squares, and diamonds) displayed in the figure are not defined; the caption states that errors are **omitted** but also that they are **obscured by the symbols**; no mention is made on the nature of such errors, nor whether the overall trend of the different data series can be regarded as showing agreement or not.

The reason I have shown these examples here is obviously not to criticise the students who have produced such plots, but rather to point out the fact that these figures will have been clear to *them*. They knew inside out the purpose of their studies and what each figure was showing in their contexts. Therefore, it is all too easy to take things for granted and assume that your reader will understand what you mean. As I hope I have just proved to you, this is far from being the case.

Arguing that the reader would have understood from the main text what the figures were all about defies the key point of figures and figure captions, namely that of standing alone and of providing useful information without expecting the reader to read the main text.

Table 2. Individual upper- and lower-row detector efficiencies for the setup without and with collimators, as obtained by two independent simulations and by using an alpha source. Unless stated otherwise, quoted errors represent a conservative systematic estimate of 5% (relative error). Calculated ratios of efficiencies for setup without and with collimators are also shown, together with the experimental ratio R obtained from the $E_{\rm p} = 151$ keV thick-target resonance yield (see text for details).

		Efficiency [%]		Efficiency ratio R
		without collimators	with collimators	(without collim./with collim.)
upper row detectors	Simulation A [17]	2.3 ± 0.1	0.20 ± 0.01	11.5 ± 0.2
	Simulation B [18]	2.3 ± 0.1	0.21 ± 0.01	11.0 ± 0.6
	α source measurements	$1.9 \pm 0.3^{(a)}$	$0.18 \pm 0.03^{(a)}$	10.4 ± 0.5
	thick-target yield			$10.1 \pm 0.2^{(b)}$
lower row detectors	Simulation A [17]	1.5 ± 0.1	0.20 ± 0.01	7.5 ± 0.4
	Simulation B [18]	1.5 ± 0.1	0.19 ± 0.01	7.9 ± 0.4
	α source measurements	$1.5 \pm 0.2^{(a)}$	$0.19 \pm 0.03^{(a)}$	7.6 ± 0.4
	thick-target yield			$8.7 \pm 0.3^{(b)}$

[a] Systematic error from the measured activity of the source.
[b] Statistical error.

Figure 7.4 Example of a complex, yet well laid-out table. The table title provides enough information to allow the reader to understand what is presented in the table.

7.3.4 What goes in a table

Tables are an extremely useful way to provide large amounts of data or results in a fairly compact form. Like figures, tables and their titles must contain some key features to guarantee readability. When preparing a table, remember that columns typically define the variables, while rows contain their values. Each column should be clearly labelled with the variable it represents, including units if applicable.

Avoid overloading the reader with an excessive number of columns; instead, consider using more tables if necessary. Variables' values (in rows) should be presented using a clear, unambiguous notation and their uncertainties (if applicable) should also be quoted together with an indication of their nature. This can be done either by adding footnote(s) to the table or by using a subscript to the value of the uncertainty to state whether the error is statistical or systematic.

7.3.5 Table titles

Tables, like figures, are also meant to stand alone. Thus, while aiming for brief titles, you should include all of the necessary details to inform the reader about what is presented in the table. Unlike figure captions (placed below the figure), table titles are placed *above* the table. An example of a well-prepared table and table title is shown in Figure 7.4.

If a table presents a large amount of information, make sure you help your reader to focus on any specific items to be compared. This can be achieved, for example, by using a bold-face type font for the items of interest so that their values stand out compared to others. Alternatively, data can be sorted in increasing or decreasing order of the most significant variable. Where applicable, state the source from where data were taken, unless they are your own.

7.4 EQUATIONS AND SYMBOLS

Equations are normally numbered sequentially as they appear in the main text, with a number in brackets on the right-hand side of the equation, so that they can be referenced throughout the text.

ISO[5] standards (see for example, [25], chapter 10) apply to the use and style of letter symbols appearing either in equations or in the main text. These are summarised below [25]:

1. Symbols representing **quantities** or **variables**. These are written in *italics*. They typically consist in single letters, from either the Latin or Greek alphabets and may contain superscripts or subscripts. Examples include: Ω (solid angle); T (temperature); E (energy). Note that if superscripts or subscripts are present, they appear in standard (Roman) fonts as in σ_{tot} (total cross section), unless they too represent a variable, as in $\Sigma_i x_i$ where i is a running number. Symbols used to represent parameters such as a and b also appear in *italics* in mathematical equations.

2. Symbols representing **units**. These, and other SI prefixes, are always used in Roman fonts, as in m/s (meters/second), μg (microgram), mL (milliliter).

3. Symbols representing **descriptive terms**. These always appear in Roman (upright) fonts. Descriptive symbols include symbols for chemical elements, such as Ar (argon), He (helium); certain mathematical symbols, such as Σ (sum of) or functions such as $\exp x$ (exponential of), dx/dt (first derivative of); and descriptive text used as superscript or subscript on quantity symbols, such as m_e (m mass, e electron).

Finally, note that different typefaces are used for a given symbol to define what type of quantity the symbol represents. For example, a quantity "A" will be written as A (*italics*) if it represents a *scalar* quantity; **A**, (**bold face**) if it represents a *vector* quantity; and as A (Roman) if it represents a unit (for example, ampere in this case).

For a comprehensive list of different types of symbols and their correct usage in text and equations, consult, for example, the *Guide for the Use of the International System of Units (SI)* [25] of the National Institute for Standards and Technology. A useful checklist for reviewing manuscripts can also be found in [25] (p. v-vi).

---

[5]The International Organization for Standardization (ISO) is an international standard-setting body composed of representatives from various national standards organizations (source: Wikipedia).

7.5 REPORTING EXPERIMENTAL RESULTS

Most scientific disciplines are based on quantitative measurements, whose results should be reported by correctly quoting their numerical values and associated uncertainties. A full discussion on different types of uncertainties and ways to analyse experimental data clearly goes beyond the scope of this book. The interested reader can refer to any of the many textbooks available on the subject and some of them are listed in the Further Reading section of this chapter.

Here, we limit our attention to recalling that experimental errors are normally stated to one leading significant figure[6] (for example, 3.4 ± 0.2 m/s) unless the significant figure is 1, in which case two significant figures should be quoted (for example, 3.42 ± 0.13 m/s). Note that the significant figure in the uncertainty dictates the number of digits used in the final result. So for example, for a distance of approximately 1200 m measured with an uncertainty of 100 m, the final result should be quoted as (1200 ± 100) m, and not (1200.5 ± 100) m, or (1183 ± 100) m.

When determining the number of significant figures in a numerical value, remember that (see for example [26]):

- All non-zero numbers are significant: 2.998×10^8 m/s has four significant figures.

- All zeroes within a number are significant: 1.380662×10^{-23} J/K has seven significant figures.

- Leading zeroes before a decimal point are not significant figures: 0.25 kV has two significant figures.

- Zeroes after a decimal point are all significant: 4.50 kg has three significant figures.

- Zeroes at the end of a number (not following a decimal point) may be significant: 270 Ω might have two or three significant figures depending on the precision of the measurement. To avoid ambiguity, consider using a *scientific notation* as 2.7×10^2 Ω or 0.27 kΩ (two significant figures) or 2.70×10^2 Ω or 0.270 kΩ (three significant figures).

7.6 APPENDICES

Appendices are rarely used in journal papers unless, occasionally, for extensive reviews. By contrast, they often appear in PhD theses or other dissertations. Yet, students are sometimes unclear as to what should go in an appendix. As

---

[6]Significant figures are defined as the figures of a number that express a magnitude to a specified degree of accuracy. More simply put, a significant figure is a reliably known digit. So, for example, the number 3.145 has 4 significant figures and indicates a precision to the third decimal place.

a rule of thumb, you should use appendices to include any supporting material that would otherwise interrupt the narrative in the main body of your report. Good items to relegate to an appendix would be, for example:

- computer codes (or parts thereof);
- complete numerical data sets or extended sets of results (often in tabular form);
- detailed descriptions of specific equipment used and/or their basic working principles;
- the full derivation of some theoretical formalism used, for example, to analyse the data;
- qualitative data sets, such as survey responses; and so on.

Appendices can therefore provide additional background information for completeness purposes as well as act as useful repositories for any additional work or detail that would otherwise be lost. So, for example, if your data analysis involved fitting a large number of experimental spectra, you would include just one or two most representative examples in the main body of your thesis and place all others in an appendix. In this way, you avoid disrupting the reader's attention from the key message, while at the same time preserving a record of all the work done.

7.7 GLOSSARY AND LISTS OF ACRONYMS

Glossaries and lists of acronyms are seldom used in scientific papers or theses. However, for very technical subjects, it may prove helpful to the reader to include, at the beginning of the thesis, a brief *glossary* to carefully explain or define key terms used throughout the work.

Similarly, if your text contains a large number of acronyms (though I would strongly discourage you from using too many, too often in the main body of your thesis) it may be useful to add a *list of acronyms* as an appendix. This will allow your readers to look for the meaning of a technical word or acronym in one place only, rather than flipping back and forth through the pages of your thesis.

The general rule for using acronyms and other abbreviations in the main body of your work is to write the terms in full the first time they occur and provide the acronym in brackets immediately after. Throughout the rest of the document, use the acronym (without any full stops between letters). For acronyms that can be used in the plural, simply add a lowercase 's' without any apostrophe (see Section 6.3.5 and Table 6.2 for rules on the use of apostrophes).

7.8 ACKNOWLEDGMENTS

Acknowledgments are optional, but often included in most PhD theses and scientific papers. Whom to acknowledge is obviously for the author(s) to decide, but at the very least it is good practice to mention the following:

- your funding sources (typically for your scholarship, if you are a PhD student, and/or the funding agency that supported your project);
- the use of facilities (for example, a major telescope, or a nuclear physics laboratory);
- the provision of services (e.g., a chemical laboratory that carried out sample preparation);
- any sharing of data (from a colleague or from available databases); and/or
- any other individual or institution that assisted or contributed to your research in some way.

Finally, for PhD theses only, you may thank colleagues for useful discussions and advice or express gratitude for the support, encouragement and motivation offered by family and friends. Just make sure you spell all names correctly!

Chapter 7: The Technical Stuff
In a nutshell...

- Titles should indicate the nature and purpose of your study.
- Figures and tables are meant to stand alone. Figure captions and table titles should contain all relevant details for the reader to be able to understand the information presented without having to read the entire paper (or chapter).
- Consider adding appendices to your thesis or dissertation for any additional material that goes beyond the key message presented in the main body of your paper or thesis.
- For very technical subjects that make extensive use of technical terms and acronyms, consider including a brief glossary or a list of acronyms for ease of reference.
- Acknowledgments typically include your funding sources and any individual or institution who has contributed in some way to your project.

EXERCISES

7.1 **Play with words**. Any time you need to come up with titles for a paper or simply for a chapter or section of your thesis, spend some time to come up with different options.

7.2 **Build your own glossary**. As a good practice throughout your study, start building your own glossary from the early stages of your PhD. All you need is a rubric with letters at the margin in alphabetical order. For any new technical term that you encounter when reading the literature, put an entry in your rubric with a brief explanation (or definition) of its meaning. You can then consult your rubric any time you need to remind yourself of the meaning of any technical terms in your discipline and, more importantly, when you need to define them in your own writing.

FURTHER READING

Bevington, P.R. and Robinson, D.K. (2003) *Data Reduction and Error Analysis for the Physical Sciences*, McGraw Hill (third edition).

Cowan, G. (1998) *Statistical Data Analysis*, Oxford University Press.

Hughes, I.G. and Hase, T.P.A. (2010) *Measurements and their Uncertainties. A practical Guide to Modern Error Analysis*, Oxford University Press.

American Chemical Society (ACS)
http://aerosol.chem.uci.edu/intranet/writing/ACS_style_guide.pdf

American Medical Society (AMA)
The AMA Style Manual
http://www.amamanualofstyle.com

American Physical Society (APS)
The APS Style Manual
http://www.apsstylemanual.org

Council of Science Editors (CSE)
The CSE Manual for Authors, Editors, and Publishers
http://www.scientificstyleandformat.org/Home.html

National Institute of Standards and Technology
The NIST Guide for the use of International System of Units
https://www.nist.gov/physical-measurement-laboratory/special-publication-811

Bureau International des Poids et Mesures
GUM: Guide to the Expression of Uncertainty in Measurements
http://www.bipm.org/en/publications/guides/gum.html

United Kingdom Accreditation Services
The Expression of Uncertainty and Confidence in Measurement
https://www.ukas.com/download/publications/publications-relating-to-laboratory-accreditation/M3003_Ed3_final.pdf

CHAPTER 8

Worked-out Examples

CONTENTS

8.1	Energy consumption in data centers	115
8.2	Colorectal cancer	120
8.3	Blainville's Beaked whales	127

MOST OF THE EXAMPLES used so far have been intentionally short and limited to highlighting some specific problems and suggesting ways to remedy them. In general, however, we need to revise and edit considerably larger amounts of text and be able to decide what to do in order to progressively improve on initial drafts.

This is especially important when working collaboratively at a research paper, or when revising entire thesis chapters. Clearly, providing full research papers or chapters to revise would go beyond the scope of this book.

Hopefully, though, a couple of excerpts from real-case examples of work submitted by students should be enough to help you appreciate the kind of revision and editing that you may have to apply to your own drafts. For each of these passages, we will focus first of all on the most pressing issues (as discussed in Chapter 4) and then on the necessary steps to improve on style and clarity (as discussed in Chapter 5). The excerpts are presented in order of increasing level of severity of issues and therefore of greater complexity of revision. The revised versions provided in this chapter are not meant to be final or perfect, but merely to illustrate the kind of improvement that one can achieve with proper revision and editing.

8.1 ENERGY CONSUMPTION IN DATA CENTERS

The first example consists in a brief introduction section of a conference paper written by a PhD student. Citations in the text appear as Arabic numerals in square brackets but the full list of references, originally included at the end, has been omitted. For easier reference and reading, paragraphs have been numbered.

Energy consumption in modern data centers

(Original version)

Introduction

1. Nowadays, the cost associated with energy consumption can be seen as one of the major concerns of data centers. This cost is sometimes non-linear with the capacity of those data centers, and it is also associated with a high amount of Carbon emission (CO_2). Some projections considering the data center energy-efficiency [1] show that the total amount of electricity consumed by data centers in the next years will be extremely high and that the associated carbon emissions would reach unprecedented levels.

2. Depending on the efficiency of the data center infrastructure, the number of watts that it requires can be from three to thirty times higher than the number of watts needed for computations [2]. And it has a high impact on the total operation costs [3], which can be over 60% of the peak load.

3. Currently, with the energy costs increasing, the focus shifts from optimizing data center resource management for pure performance to optimize it for energy efficiency while maintaining the services' performance [4]. In this scenario, strategies that are aware of energy consumption are gaining attention of the academy and industry, since they can fulfil the promise of executing applications that can be tuned to consume less energy [5].

4. The question is how to increase the energy-efficient of the whole data center without sacrificing Quality-of-Service (QoS) requirements, both for economical reasons and for making the IT environment sustainable [6]. Answering that question is difficult since there are many variables that contribute to the power consumption of a computing resource. For instance, the power consumption of a node does not only depend on its architecture or on the application it is running, but it also depends on its position in the data center and on the temperature of the data center [7]. Thus, efficient resource manager has an important role in a data center to help reducing the energy consumption.

5. Many energy-efficient computing approaches focus on single-objective optimizations, without considering the QoS parameters. Energy-saving schemes that result in too much degradation of the system performance or in violations of Service-Level Agreements (SLAs) parameters would eventually cause the users to move to another provider. Therefore, there is a need to reach a balance between the energy savings and the costs incurred by these savings in the execution of the applications.

continues...

Worked-out Examples ■ 117

> ...continued
>
> **6.** In that context, cloud computing is gaining popularity since it can help companies to reduce costs and carbon footprint, usually distributing service execution across distributed data centers. To support a large number of consumers or to decentralize management, clouds can be combined, forming a federated cloud environment. A federated cloud can move services and tasks among clouds in order to achieve their goals. Those goals are usually described as QoS metrics, such as minimum execution time, minimum price, availability, minimum power consumption and minimum network latency, among others. Federated clouds are an elegant solution to avoid overprovisioning, thus reducing the operational costs in an average load situation, while still being able to give QoS guarantees to the users. However, in order to improve the cloud efficiency it is necessary to consider the efficiency of resource allocation in a dynamic scenario.

This excerpt offers a nice example of a well-organised text, with a large amount of information presented in a meaningful sequence. The structure therefore is not the main issue for this example. The text, however, lacks focus somehow; it contains a fair amount of jargon; and it is occasionally difficult to read. In some places, both clarity and style can be improved by avoiding repetitions and removing unnecessary details, which could be relegated to later sections and further expanded upon. Finally, the overall length can be reduced to make the text crisper and easier to read.

For drafts such as this one, where the structure is already well in place and only minor edits are needed, the best way to proceed consists in highlighting the key ideas that we wish to retain in each sentence or paragraph and grouping them together, leaving out any unnecessary repetitions. Before proceeding with the revision, let us examine more closely each paragraph in turn.

- **Paragraph 1.** This paragraph contains a couple of key ideas that are important to retain, namely the **cost of energy consumption increases non-linearly with capacity** and **carbon emissions will reach unprecedented levels** in the near future.

- **Paragraph 2.** Similarly, this paragraph highlights the link between data centers' efficiency and their total cost, which can reach a significant fraction of the peak load.

- **Paragraph 3.** Here, the key idea is around a **focus shift... in resource management** towards **strategies** that can help to **consume less energy**. Note how the first sentence is long-winded, while the second could be rephrased for improved clarity and better emphasis.

- **Paragraph 4.** This paragraph gradually moves to a specific problem (The question is...) of data center operations and how these should be managed differently in order to help reducing the energy consumption, without sacrificing the Quality of Service. Note how some of the information (For instance, the power consumption...) presented here is perhaps too detailed for the purpose of this introduction and could be moved to a later section.

- **Paragraph 5.** Here, the author warns about the possible danger for users to move to another provider unless a balance can be found between energy-saving schemes and their impact on execution of applications that respects Service-Level Agreements (SLAs) parameters. These are therefore key ideas to retain in the revised version.

- **Paragraph 6.** Finally, the introduction closes with a paragraph that offers a possible solution to the problem presented earlier and points to a specific requirement, namely of allocating resources in a dynamic scenario through cloud computing. Note, however, how the paragraph leaves the reader somewhat mid-air, without a clear indication of where the text is going next. Ideally, this could be improved by adding, for example, some information about the paper and what the reader can expect to find in it, or what the future directions are in this research context.

The following focus box offers an example of a possible revision.

Energy consumption in modern data centers

(Revised version)

Introduction

1. The cost associated with energy consumption of modern data centers has increased non-linearly with data centers' capacity. Projections based on current energy-efficiency indicators predict that the total amount of energy consumed by data centers over the next few years will lead to unprecedented levels of CO_2 emissions. Depending on their efficiency, data centers' infrastructures will require from three to thirty times more power than is needed for carrying out computations, with a consequent increase in operational costs of over 60% of the peak load.

continues...

...continued

2. Unsurprisingly, current focus has thus shifted from managing resources for pure performance to optimizing resources for energy efficiency, while at the same time maintaining the Quality of Service (QoS). Both academia and industry are turning their attention to strategies that promise to execute energy-efficient applications and guarantee the long-term sustainability of the IT environment. However, energy-saving schemes that result in too much degradation of the system performance or in violations of Service-Level Agreements (SLAs) parameters would eventually cause users to move to a different provider. Thus, the efficient management of resources that aims to reduce energy consumption poses severe challenges and requires a fine balancing between energy savings and the costs incurred in the execution of applications.

3. Within this context, cloud computing (i.e., the distribution of service execution across different data centers) has recently gained popularity thanks to its reduced cost and carbon footprint. To support a large number of consumers or to decentralize management, clouds can be combined into a federated cloud environment, where services and tasks can be shared or moved among clouds in order to achieve specific goals. These goals, referred to as QoS metrics, include minimum execution time, minimum price, availability, minimum power consumption and minimum network latency, among others. Federated clouds provide an elegant solution to avoid over-provisioning, to reduce the operational costs in an average load situation, and to guarantee QoS to their users. However, resource allocations must be considered in a dynamic scenario in order to improve the cloud efficiency.

Note how this revised version maintains the original structure while at the same time shortening the text (from about 520 words to 330; about a 35% cut) by avoiding unnecessary details and repetitions. In particular, note how, following the upcoming template presented in Table 9.1, the first paragraph provides an overview to the topic and offers some key background information to place the research in context.

This is then followed by a "development paragraph" that further elaborates on the problem sketched previously and highlights areas of current research interest. This second paragraph closes with the challenges faced, which provides a smooth transition onto the final, closing paragraph.

Finally, the third paragraph hints at a possible solution of the research problem discussed previously. However, unlike the template in Table 9.1, this paragraph does not close with an overview of the paper nor of the work done by the present study to address this research issue. This can easily be added if needed for the type of publication for which this introduction was written.

8.2 COLORECTAL CANCER

The second example consists in two sections (*Introduction* and *Biological background*) from a scientific report written by a PhD student and aimed at an audience largely familiar with the topic. For easier reference and reading, paragraphs have been numbered, while bibliographical references (appearing as Roman numeral superscripts in the original text) have been removed.

STUDY OF THE MECHANICAL PROPERTIES OF 3D TISSUE STRUCTURES USING ATOMIC FORCE MICROSCOPY (AFM)

(Original version)

Introduction

1. It is becoming increasingly clear that the mechanical properties of cells and tissue change during the progression of cancer. Efforts have been made to study the mechanical characteristics of cells and tissues, using methodologies that allow study of small forces (pN-nN). Such techniques include traction force microscopy, embedded particle tracking, micropipette aspiration, optical and acoustic stretchers, atomic force microscopy, optical and magnetic tweezers, and microneedles. This wide range of methodologies is generally limited to single cell and adhesion experiments and is always capable of providing quantitative information. On the other hand, clinical elastography techniques such as ultrasound elastography do not provide cellular resolution. We therefore aim to develop novel methods to measure the mechanics of emerging three-dimensional organoid structures, attempting to bridge the gap between single cell and whole tissue. An important issue is that organoids are grown in a matrix, which itself has physical properties that need to be considered, creating an additional necessity for novel approaches, including techniques that do not require direct contact with the sample.

2. Here we demonstrate the application of atomic force microscopy (AFM) to three-dimensional organoid systems. AFM uses the deflection of a microscopic cantilever to measure local surface properties of a sample with high resolution. When used in force spectroscopy mode, it provides information about the local stiffness (Young's modulus) and adhesiveness of a sample. We explore how careful sample preparation and system parameters can be optimised to measure local mechanical properties of cells within such a 3D system. We conclude that AFM is a reliable, quantitative method to measure local mechanical properties, though due to limitations of the technique, we will continue to explore other methods as well throughout the remainder of this PhD project.

continues...

...continued

Biological background

3. Colorectal cancer is the second most common cause of cancer deaths in the developed world, accounting for 12% of all cancer deaths in Europe, and almost 700,000 total deaths in the world every year. The incidence of colorectal cancer is predicted to increase in the near future, as lifespans increase and more people adopt Western lifestyles and diets.

4. In the digestive tract, cells originate from stem cells in the crypt base and migrate upwards towards the gut lumen where they are shed within 3 to 5 days, ensuring continual removal of cells. When mutations occur that affect migration, this process is inhibited or slowed down, which can result in the development of cancer. In over 90% of colorectal cancers, mutations in the adenomatous polyposis coli (Apc) gene are present. The APC protein is linked to proliferation, differentiation and migration, consistent with changes in all these processes in cancer.

5. The biochemical and cell biological consequences of mutations in Apc have been investigated in detail, whereas their effect on cell and tissue mechanics is completely unexplored. Generally, tumours are stiffer than the surrounding healthy tissue, a property that has been attributed to changes in the extracellular matrix (ECM). Additionally, AFM has revealed that single tumour cells are softer than healthy cells in tissue. The aim of this project (ESR10) is to examine the mechanical properties of cells in gut epithelial tissue using gut organoids as a tissue model. Organoids represent only the epithelial layer, the tissue where colorectal cancer originates, but maintain a three-dimensional spatial organisation that mimics the situation in situ. The effect of specific Apc mutations on mechanical properties of individual cells and the consequence for three-dimensional organisation will be also investigated.

Unlike the previous example, the main problem of this excerpt is that its structure lacks coherence and some information is not presented in the right sequence. This issue needs to be tackled first before proceeding to other less pressing ones. These include some technical terms being used without explanation or definition as well as stylistic issues. Also, the aim of the study is not immediately clear and this too should be rectified. Again, let us inspect each paragraph in turn before making any changes.

- **Paragraph 1**. Here the writer provides background information to place the study in context. The first few sentences explain that **mechanical properties of cells and tissue change during cancer** and that several different approaches are used to study what these properties are. The writer also tries to compare advantages and disadvantages of different approaches as indicated by the phrases ... **generally limited to single cell, always capable of providing quantitative information, do not provide cellular resolution**. However, these are not spelled out sufficiently well. As a result, even though the writer states that the aim of the study is to **develop novel methods... to bridge the gap between single cell and whole tissue**, it is not clear why this is an important issue to address.

The final sentence is also problematic. It presents new ideas, namely that **organoids are grown in a matrix** and that the matrix itself **has physical properties that need to be considered**. However, this information simply comes in too late – at the conclusion of the paragraph – and does not fully explain why novel approaches are needed that **do not require direct contact with the sample**. As a result, the final sentence appears slightly disconnected from what presented so far and reads more like an add-on afterthought. Other, minor issues are more of stylistic nature and regard repetitions, the use of words never defined (such as **organoids** and **matrix**), and empty phrases such as **It is becoming increasingly clear that**.

- **Paragraph 2**. In this paragraph, the writer briefly describes the method (**atomic force microscopy**) used in the study and states how **careful sample preparation and system parameters can be optimised to measure local mechanical properties of cells within such a 3D system**. Here, a 3D system likely refers to the **three-dimensional organoid structures** mentioned in the last-but-one sentence of the first paragraph. However, this reference is too far away and therefore unclear. Finally, **limitations of the technique** are mentioned but, without stating what these are, it is difficult to appreciate why other methods should be explored.

- **Paragraph 3**. This paragraph offers some interesting factual information about the incidence of colorectal cancer. Even though there is no explicit mention to *why* it is important to study its causes and possibly to find a cure, this type of information is likely to attract the reader's attention. Ideally, it should be placed earlier on, possibly at the beginning of the first paragraph in the introduction.

- **Paragraph 4**. Here, the writer provides background information about the way in which cancer can develop in the digestive tract. Some technical terms (**crypt base, gut lumen, adenomatous polyposis coli**) are used without further explanation and may remain obscure to a non-expert audience. On the other hand, their use may be appropriate for an audience familiar with the topic, as the one likely addressed by this report. If the report caters for a mixed audience, some words of explanation, for example in a footnote, would be useful.

- **Paragraph 5**. This paragraph contains important information needed to understand much of what presented previously and so it should ideally appear earlier in the text. This is the case for information given to explain some key aspects of the topic, such as: **single tumour cells are softer than healthy cells in tissue**, or to define terms such as **organoids**. Similarly, the sentence **The aim of this project is to examine mechanical properties of cells in gut epithelial tissue using gut organoids as a tissue model** should appear earlier (possibly in the *Introduction*) to provide the reader with a better understanding of the context and purpose of this study.

Finally, while the sentence **their effect on cell and tissue mechanics is completely unexplored** provides an excellent motivation for the study, it appears in mid paragraph and loses emphasis as a result.

Having explored in greater detail the content and partly the function of each paragraph, we can better appreciate how the main issue of this second draft example is essentially a problem of structure. The first step in its revision should then aim at re-organising information presented. The following focus box shows the revision made by the student following my feedback.

STUDY OF THE MECHANICAL PROPERTIES OF 3D TISSUE STRUCTURES USING ATOMIC FORCE MICROSCOPY

(First revision)

Introduction

1. Colorectal cancer, the second most common cause of cancer deaths in the developed world, accounts for 12% of all cancer deaths in Europe and almost 700,000 total deaths in the world every year. The incidence of colorectal cancer will likely increase in the near future, as lifespans increase and more people adopt Western lifestyles and diets.

2. During the progression of cancer, the mechanical properties of cells and tissue change. Quantification of these changes has either been focussed on the mechanical characteristics of cells or on those of tissues. For cells, methodologies are required that allow measurement of small forces (pN-nN). Such techniques include traction force microscopy, embedded particle tracking, micropipette aspiration, optical and acoustic stretchers, atomic force microscopy, optical and magnetic tweezers, and micro-needles. This wide range of methodologies is typically limited to single cell and adhesion experiments and is usually capable of providing quantitative information. For example, atomic force microscopy (AFM) studies have revealed that single tumour cells are softer than healthy cells in tissue.

continues...

...continued

3. On the other hand, clinical elastography techniques such as ultrasound elastography inform on the mechanical properties of tissue, though they do not provide cellular resolution. Generally, tumours are stiffer than the surrounding healthy tissue, a property that has been attributed to changes in the extracellular matrix (ECM).

4. In an attempt to bridge the gap between single cell and whole tissue, we aim to develop novel methods to measure the mechanics of emerging organoid structures. An organoid is a three-dimensional group of cells that mimics a specific tissue or organ in a laboratory environment. It contains different cell types and mimics the behaviour and organisation of *in vivo* tissue. Organoids provide a useful tool for studying developmental and cancer biology in a more physiological relevant setting than monolayer cell culture, as well as providing a high-throughput tool for drug discovery. For gut organoids, the tissue where colorectal cancer originates, the gut epithelium, can be studied in isolation, while maintaining a three-dimensional spatial organisation.

5. Organoids are grown in a 3D matrix called Matrigel which provides the correct chemical and mechanical environment for the structures to form. This introduces an extra variable, as it has its own physical properties that need to be considered. Therefore novel approaches are needed, such as techniques that do not require direct contact with the sample, or optimised sample preparation that allow access to the structures of interest.

6. In this report, we focus the application of atomic force microscopy (AFM) to three-dimensional organoid systems. AFM uses the deflection of a microscopic cantilever to measure local surface properties of a sample with high resolution. When in force spectroscopy mode, AFM provides information about the local stiffness (Young's modulus) and adhesiveness of a sample. We explore how careful sample preparation and system parameters can be optimised to measure local mechanical properties of cells within a 3D structure. We conclude that AFM is a reliable, quantitative method to measure local mechanical properties. However, the technique is limited to measurements on a cellular scale, so we will continue to explore complementary methods throughout the remainder of the PhD project.

continues...

> **3D tissue models for colorectal cancer**
>
> **7.** In the digestive tract, cells originate from stem cells in the crypt base and migrate upwards towards the gut lumen where they are shed within 3 to 5 days. This process ensures continual removal of cells. When mutations occur that affect migration, this process is inhibited or slowed down, which can result in the development of cancer. In over 90% of colorectal cancers, mutations in the adenomatous polyposis coli (Apc) gene are present. The APC protein is linked to proliferation, differentiation and migration, consistent with changes in all these processes in cancer. The biochemical and cell biological consequences of mutations in Apc have been investigated in detail, whereas their effect on cell and tissue mechanics is completely unexplored.
>
> **8.** We propose using 3D tissue culture to study the mechanical effects of Apc mutations. Gut organoids mimic intestinal epithelial tissue; these "mini-guts" can be grown in a 3D matrix called Matrigel starting from isolated intestinal crypts or individual stem cells. Wild type gut organoids form crypt-like structures with the same cell organisation and behaviour as crypts in tissue. Organoids grown from Apc-mutant crypts, lack cellular organisation and grow into round spheroids (cysts).

Having improved the overall structure, we can now focus on other issues to improve on style and clarity following the advice presented in Chapter 5. Again, it is helpful to inspect each paragraph in turn first.

- **Paragraph 1.** This paragraph now provides an excellent opening to the *Introduction* section. The language is clear and easy to understand. Notice the use of strong verbs such as **accounts**, **increase**, and **adopt**. No change is necessary here.

- **Paragraphs 2 and 3.** These two paragraphs together now offer a better attempt at explaining advantages and disadvantages of different techniques. Specifically, some techniques (paragraph 2) provide quantitative information, but only on individual cells, while others (paragraph 3) provide information on mechanical properties of tissues as a whole, but lack cellular resolution. However, these paragraphs could be revised slightly for improved clarity and added emphasis.

- **Paragraph 4.** The specific aim of the study becomes more explicit in this paragraph and its relevance more obvious. The paragraph also contains a useful definition of what an organoid is and explains why organoids are useful for this type of study.

- **Paragraph 5**. This paragraph could be removed from the introduction and used later to expand more on the method. Potentially, the only information worth retaining refers to the fact that an organoid **has its own physical properties that need to be considered**. However, it should be possible to incorporate this piece of information in the previous paragraph.

- **Paragraph 6**. As in the original version, this paragraph provides a concise overview of the aim of the study, the approach followed, and the results obtained. Unlike its previous version, it now explicitly states the limitation of the approach used in this study.

- **Paragraph 7**. Moving on to a new section, the writer now provides more background information of how colorectal cancer develops and presents the main gap in current research (**their effect on cell and tissue mechanics is completely unexplored**).

- **Paragraph 8**. The paragraph re-states the aim of the study, but this time in slightly more specific terms (**using 3D tissue culture** and **to study the mechanical effects of** Apc **mutations**). It also provides further information on the methods used.

A possible revision (limited to the first few paragraphs only) is shown below. Note how some paragraphs have now been split, while others have been merged to better signpost concepts. Parallel structure has been used where necessary to improve on clarity, as indicated in boldface.

STUDY OF THE MECHANICAL PROPERTIES OF 3D TISSUE STRUCTURES USING ATOMIC FORCE MICROSCOPY

(Second revision)

Introduction

1. Colorectal cancer, the second most common cause of cancer deaths in the developed world, accounts for 12% of all cancer deaths in Europe and almost 700,000 total deaths in the world every year. The incidence of colorectal cancer will likely increase in the near future, as lifespans increase and more people adopt Western lifestyles and diets.

2. During the progression of cancer, the mechanical properties of cells and tissue change. While several techniques have been developed to quantify these changes, they can reveal properties of either single cells or whole tissues, but not both at the same time (**better emphasis**).

continues...

continued...

3. For cells, methods are required that allow to measure small forces (pico- to nano-Newton). Such techniques include traction force microscopy, embedded particle tracking, micropipette aspiration, optical and acoustic stretchers, atomic force microscopy, optical and magnetic tweezers, and micro-needles. These techniques can provide quantitative information – for example, atomic force microscopy (AFM) studies have revealed that single tumour cells are softer than healthy cells in tissue – but they are limited to single cells and to adhesion experiments.

4. For tissues, clinical elastography techniques such as ultrasound elastography can inform on the mechanical properties of whole tissues, but do not provide cellular resolution. Generally, such techniques have revealed that tumours are stiffer than the surrounding healthy tissue, a property attributed to changes in the extracellular matrix (ECM).

5. In an attempt to bridge the diagnostic gap between single cell and whole tissue, we aim to develop novel methods to measure the mechanics of emerging *organoid structures*: three-dimensional groups of cells that mimic a specific tissue or organ in a laboratory environment. Organoids contain different cell types and resemble the behaviour and organisation of *in vivo* tissue. They therefore offer an exciting opportunity to study developmental and cancer biology in a more physiological relevant setting than monolayer cell culture, as well as a high-throughput tool for drug discovery.

6. However, since organoids are grown in a 3D matrix that provides the correct chemical and mechanical environment for structures to form, the matrix's physical properties also have to be taken into account (**added emphasis**). Thus, novel approaches are needed, that either do not require direct contact with the sample, or that require optimised sample preparation to allow access to the structures of interest (**parallel structure**).

8.3 BLAINVILLE'S BEAKED WHALES

The final example presented here is an excerpt from the introductory chapter of a PhD thesis on a study on communication among Blainville's beaked whales. The thesis was organised around four main chapters each devoted to a specific aspect of the research and intended to stand alone for publication as individual papers. The introduction and conclusion chapters were meant to provide an overarching framework for the four individual chapters. The student got in touch with me three months before the submission of her thesis, asking for additional support with her write-up.

As most chapters had already been drafted, except for the *Introduction* and the *Conclusion* chapters, we decided to work on the Introduction together, so that she could put into practice all the necessary writing steps, as presented in this book, and later adopt the same framework for writing her final chapter.

The first task I gave her was then to prepare a mind map of the content of the *Introduction* chapter and to convert it into a structured layout (Section 3.4). This would provide a canvas for both of us to work on. The layout she prepared is shown in the following focus box.

CHAPTER ONE - GENERAL INTRODUCTION

(Layout)

1.1 Animal Communication
- Types and functions
- Current literature debate on information
- Cues and evolution
- Eavesdropping
- Signals and constraints
- Mother/calf

1.2 Communication in cetaceans
- Types and functions
- Cues
- Mother/calf – pinniped example

1.3 Cetacean social structure
- Stable
- Fission-fusion

1.4 Blainville's beaked whales
- What we know now
- Why we need to know more
- Constraints

To show the progression of her work, and to demonstrate the level of improvement that can be achieved through revision, I will present in order an early draft, its first revision (following feedback), and the final submitted version. Because of space constraints, I reproduce here only a minor excerpt of the whole chapter.

CHAPTER ONE - GENERAL INTRODUCTION

(Early Draft)

1. Blainville's beaked whale research has had more funding in the last decade than ever before, following a number of mass stranding events coincident with naval exercises (Balcomb & Claridge, 2001; Van Bree & Kristensen, 1974; Cox et al., 2006; Evans & England, 2001; Fernández et al. 2005; Frantzis, 1998; Jepson et al., 2003; Simmonds & Lopez-Juraco, 1991). The focus of recent studies has been directed at beaked whales' response to sound sources, in the form of documenting their movement away from potentially damaging sounds (Allen et al., 2014; Deruiter et al., 2013; Tyack et al., 2011). These studies have not however investigated the potential effects of sound on beaked whale communication systems, or provided any insight into the whales' communication. This is in contrast to earlier studies on the effects of noise on whales, specifically ship noise, that concentrated on the effects the noise had on the whales communication systems, concluding the noise was masking marine mammal communication sounds (Payne & Webb, 1971).

2. Communication is ubiquitous in the animal kingdom. I introduce some of the core concepts and their evolution in the realm of modern thought of animal communication, and its on-going debate. I also introduce possible synthesis between social structures and communication systems, and finally present my study species and what is known to date.

1.1 ANIMAL COMMUNICATION

3. Animal communication occurs when a signal deliberately given by a sender alters the behaviour of a receiver (Wilson, 1975). By contrast, cues are obligate and non intentional, and their emission does not benefit the producer (Seeley, 1989), for example a mosquito can detect the cue of CO_2 from a mammal upwind. Alternatively, some obligate cues can benefit the producer, for example the quality of a male lion (*Panthera leo*) distinguished from his mane (Schaller, 1972), or whether or not a female Asian elephant (*Elaphus maximus*) is sexually receptive, detected in her urine by male elephants (Rasmussen et al., 1982). Cues have been shown to evolve to become signals through ritualisation (Tinbergen, 1952), such as the canine behaviour of baring teeth. This cue, intimating the opponent to stay away, evolved from dogs baring their teeth immediately before biting, to get their lips out of the way (Krebs & Dawkins, 1984).

continues...

continued...

4. Signals can vary in form, meaning and benefit to the signaller. The Puerto Rican crested anole (*Anolis cristatellus*) displays its ability to escape a predator from the number of push-ups it can perform during certain predation events, which also acts as a signal to potential mates on the individual's quality (Leal, 1999). Chimpanzees that groom other individuals are in turn given food sharing privileges with that individual (de Waal, 1989). Birds and mammals produce a variety of acoustic signals that can be beneficial for conspecifics outside of visual range. Vervet monkeys (*Chlorocebus pygerythrus*) produce acoustic signals specific to the type of predator approaching, for example a "cough" call indicates an aerial predator, such as an eagle is approaching and therefore hiding amongst flora should be sought as a response (Seyfarth et al., 1980).

5. In the current literature debate on whether animal signals provide information (Seyfarth et al., 2010), or merely influence the receiver, Rendall & Owren (2013) hold the vervet monkey research as a catalyst to the direction that animal communication research underwent. They argue that too many comparisons to human language were and are still being made in primate research, and for other animals, there is too great a focus on the information that animal signals convey. I strongly agree with their argument that there is no encoding within signals of certain information such as size and in some cases individual identity that naturally occur from a large vocal apparatus or characteristics of that apparatus (Fitch & Hauser, 1995; Rendall et al., 1998). However I do not know whether comparisons to language have limited the field in its range of questions (Author, year).

6. In regard to marine mammals, information is passed often unintentionally through eavesdropping, whereby a receiver exploits the sound made by another animal (Bradbury & Vehrencamp, 1998). Eavesdropping can be beneficial to a predator, as well as to prey. For example, fish-eating killer whales - who presumably cannot be detected acoustically by their prey - produce many more vocalisations during foraging attempts than mammal-eating killer whales (Barrett-Lennard et al., 1996). Likewise, eight moth families have evolved ears sensitive to ultrasonic calls used by bats to locate the moths and prey upon them (Miller & Surlykke, 2001). It is perhaps for this reason that bats do not echolocate in good lighting, for example a full moon night (Bell, 1985), but only when they need to.

7. Eavesdropping is particularly useful for animals that spend the majority of their time in the dark and communicate acoustically, such as bats and deep diving cetaceans. Wild rough-toothed dolphins (*Steno bredanensis*) use eavesdropping when travelling synchronously in a group, hearing the echolocation clicks and echoes from other members in the group (Goetz et al., 2006). This may be an energy-saving mechanism, or else one that aids in allowing fewer echoes to be processed without the back scatter that would occur if every dolphin was echolocating in the same direction.

From this early draft it is clear that the student had read extensively and was very familiar with her research topic. The information, however, is not properly structured and the text is overloaded with examples and difficult to read. Structure is not the only issue, though. Other issues include: lack of clarity, colloquialisms, excessive level of detail, grammar, punctuation, ambiguous referencing, and inappropriate content. Luckily for me, this is a perfect example to discuss in a book like this! So, let us examine each paragraph in turn before considering the steps required to revise the text.

- **Paragraph 1**. The chapter opens with a nice sentence that immediately draws the reader's attention and implicitly addresses a possible motivation for this study: **a number of** unexplained **mass strand events**. The sentence is supported by extensive literature citations (here given in the Harvard referencing format – see Section 6.5.2). Yet, the paragraph fails to provide a clear overview of the current status of research in this area. Presumably, the intended message here is that sound can have an effect on both the way in which whales *behave* and the way in which whales *communicate* and while some studies have focussed on the former (see second sentence), not much is known about the latter (see third sentence). This distinction, however, is not immediately obvious because the language used is not clear (the phrase **in the form of documenting their movement...** is used here to imply *behaviour*). Clarity is further compromised by the last sentence (**This is in contrast to earlier studies...** and by the use of the word **noise** as opposed to **sound** used previously, which makes the reader wonder whether there is an implicit difference between the two that one should be aware of.

- **Paragraph 2**. This paragraph opens with an excellent sentence: it is short, clear, and to the point. However, the sentence seems misplaced. A transition link is missing between this and the previous paragraph that explicitly mentions the motivation behind the study and the open question(s) that it intends to address. In addition, the paragraph becomes colloquial by abruptly switching to the use of the first personal pronoun *I*. The language is either vague (**some of the core concepts**), pompous (**realm of modern thought**), or obscure (**possible synthesis between social structures and communication systems**). As a result, it lacks focus as well as a coherent level of formality.

- **Paragraph 3**. Again, the paragraph opens with a nice short sentence that defines animal communication. The sentence is then followed by one that introduces a different type of communication and alerts the reader to a potentially important distinction between *signals* and *cues*. However, the many examples provided distract the reader, who has to wait until the next paragraph before being able to appreciate the full difference between signals and cues. A better approach here would be to first introduce different types of communication and then provide examples for each. Albeit minor at this stage, other issues concern a lack of parallelism in: **the quality of a male lion...** (noun + preposition + noun) and **or whether or not a female Asian elephant is sexually receptive** (conjunctions + adjectives + noun + verb); as well as an awkward use of grammar where **detected** does not seem to refer to anything.

- **Paragraph 4**. Again, the key message of the paragraph seems lost among the many examples. Also, it is unclear how they relate to the central focus of the study, namely communication among whales. Even though the writer offers a general overview of different communication strategies, the purpose of giving such an overview is not made explicit. On a minor issue, a missing hyphen between **food sharing** makes the sentence awkward to read and understand. Note also the awkward grammar at the end of the paragraph, where the last sentence: **a cough call indicates an aerial predator, such as an eagle is approaching and therefore hiding amongst flora should be sought as a response** would become clearer if rephrased as, for example: **a cough call indicates an aerial predator, such as an approaching eagle, and prompts conspecifics to hide amongst flora.**

- **Paragraph 5**. This paragraph focuses on current debates in the literature, but it is misplaced in an introduction and should be deferred to a discussion chapter. There, a more general discussion can be presented that also reflects on the outcome from the present study. Note also how the language used, **I strongly agree** and **I do not know**, is either too strong or too colloquial for academic writing and should be revised (Chapter 5).

- **Paragraph 6**. Here, yet another type of communication (**eavesdropping**) is introduced. However, without any previous reference to it, and with the added overload of further examples, the reader may feel dragged along without knowing where (s)he is heading. In other words, the reader may wonder about *why* each example is presented and *what* is important about it. Indeed, if presenting several different types of communications is needed, it would be far better to mention all of them first and then provide examples for each explaining the correlation among them. Note also how the final sentence (**It is perhaps for this reason that...**) seems to draw a conclusion from the previous one, but the connection is not immediately clear (or not to me at least!).

- **Paragraph 7**. This paragraph provides further examples on eavesdropping as a way of communicating and although one of the examples is more closely related to the core topic of the study (**deep diving cetaceans**), the paragraph lacks structure (like others before) and its key message remains somewhat vague.

After discussing these issues with the student and suggesting ways to improve her draft, I received a revised version, as given in the next focus box.

Chapter One - General Introduction

(Revised version)

1. Blainville's beaked whales (*Mesoplodon densirostris*) have been the focus of more dedicated research in the last decade than ever before, following a number of mass stranding events coincident with military naval exercises (Balcomb & Claridge, 2001; Van Bree & Kristensen, 1974; Cox et al., 2006; Evans & England, 2001; Fernndez et al., 2005; Frantzis, 1998; Jepson et al., 2003; Simmonds & Lopez-Juraco, 1991). Recent studies have focused primarily on beaked whales' behavioural response to anthropogenic noise sources, to quantify their movement away from potentially damaging sounds (Allen et al., 2014; Deruiter et al., 2013; Tyack et al., 2011). However, the effects of anthropogenic sound on beaked whale communication systems have not been investigated. This is in contrast to earlier studies on the effects of anthropogenic noise on marine mammal communication, specifically ship noise. Ship noise reduces the distances over which blue (*Balaenoptera musculus*) and fin (*Balaenoptera physalus*) whales are able to communicate (Payne & Webb, 1971), and causes North (*Eubalaena glacialis*) and South Atlantic right whales (*Eubalaena australis*) to alter their call frequency (Parks et al., 2007). Despite the recent interest in beaked whale biology, as yet there is still no insight into beaked whale communication. In order to fully understand the effect sound has on these whales, understanding their communication system is paramount.

2. In this introductory chapter, I will present an overview of the way in which animals communicate, introducing some of the core concepts. I will also discuss how social structures of animals relate to their communication systems, and introduce the study species, Blainville's beaked whales, and what is known about them to date. Finally, I will provide an overview of the thesis.

1.1 Animal Communication

3. Communication is ubiquitous in the animal kingdom. Animal communication occurs either by non-voluntary cues, or by signals. Cues are traits or behaviours that are obligate and non-intentional, and their emission does not generally benefit the producer (Seeley, 1989); for example, a mosquito can detect the cue of CO_2 from a mammal upwind, providing the mosquito with some location information for foraging on the mammal, that the mammal surely does not benefit from.

continues...

continued...

4. Yet some obligate cues can benefit the producer; for example, the fitness of a male lion (*Panthera leo*) distinguished from his mane (Schaller, 1972), or the sexual receptiveness of a female Asian elephant (*Elaphas maximus*) detected in her urine by male elephants (Rasmussen et al., 1982). In some cases, cues have been shown to evolve to become signals through ritualisation (Tinbergen, 1952), such as the canine behaviour of baring teeth. This cue, intimating the opponent to stay away, evolved from dogs baring their teeth immediately before biting, to get their lips out of the way (Krebs & Dawkins, 1984).

5. Signals, in contrast to cues, are deliberately given by a sender, and can alter the behaviour of a receiver (Wilson, 1975). Signals can vary in form, function and benefit to the signaller. Forms include visual, acoustic, tactile and chemical. Functions include reproductive success through sexual advertisement and mate attraction, alarm calls to alert conspecifics to the presence of a predator, conflict resolution, and individual identification, used commonly in mother-offspring recognition and maintaining group cohesion. Additionally, bats and cetaceans use acoustic signals to detect and localise prey. I provide some examples of combinations of forms and functions below.

6. Visual signals can be used to attract females, and also deter predators or competing males by providing an indication of fitness and dominance. The Puerto Rican crested anole (*Anolis cristatellus*) displays its ability to escape a predator by the number of push-ups it can perform during certain predation events. This display benefits the anole by also acting as a signal to potential mates on the individual's quality (Leal, 1999). Bowerbirds (family *Ptilonorhynchidae*) have evolved their sexual advertisement away from their own plumage, to the elaborate decoration of the bowers they build. This evolution has enabled bower birds to indicate through the bower quality, their age, experience and dominance to the females, as in order to produce the most elaborate bower, they will have had to steal and possibly destroy another male's bower (Pruett-Jones & Pruett-Jones, 1994). Chemical signals are the oldest form of communication and themselves vary widely in form and function. A common use of chemical signalling is territorial defence through scent marking as seen in dogs. Scent marking can also communicate dominance to other males, shown in wild brown bears (*Ursus arctos*) (Clapham et al., 2012). Tactile signals have been studied extensively in primates, and can provide benefit to the giver (sender) of the signal. Chimpanzees that groom other individuals are in turn given the benefit of food-sharing privileges with that individual (de Waal, 1989).

Let us again inspect each paragraph in turn.

- **Paragraph 1**. The opening sentence supported by its citations has remained in place. Now the second and third sentences provide a clearer distinction between the effect of sound on both behaviour and communication. However, the following sentence **This is in contrast to earlier studies...** seems to contradict what just was stated and sets the reader off track. In other words, the writer first states that **the effects of anthropogenic sound on beaked whale communication systems have not been investigated** but then mentions previous **studies on the effect of anthropogenic noise on marine mammal communication**. It was not until I specifically queried the student on this point that she explained that, yes, there had been previous studies on marine mammal communication, but not on beaked whales (also marine mammals!). The distinction between the two, however, had not been made clear in the text.

  On a positive note, the links missing in the early draft are now at least sketched in the final two sentences. Specifically, the first one highlights the gap in current research (**Despite the interest... there is still no insight into beaked whale communication**); while the second hints at a motivation by stating that **understanding their communication system is paramount**, albeit without explaining *why*.

- **Paragraph 2**. This paragraph now provides a nice and concise overview of what the reader can expect to find in the chapter. Admittedly, the language may sound too colloquial to some, but the use of the first personal pronoun in this case (in my view at least) is not entirely out of place.

- **Paragraph 3**. The opening sentence of this paragraph is now in its proper place. The following sentence introduces the concepts of *cues* and *signals* and these are further developed, in this order, over the next two paragraphs, each supported by specific examples of cues and signals, respectively, thus creating a parallel structure that is easier to follow.

- **Paragraph 4**. The paragraph offers examples that support the key idea, namely that cues can benefit the producer. Also, the previous lack of parallelism has now been fixed here as: **the fitness of a male lion...** (noun + preposition + noun) and **the sexual receptiveness of a female Asian elephant...** (noun + preposition + noun), making the sentence clearer. Also note how **detected** now clearly refers to **sexual receptiveness**.

- **Paragraph 5**. Here, the concept of signals is fully explained so that the paragraph offers a nice overview to help the reader make sense of the examples to come. Note how the final sentence in this paragraph takes the reader by the hand (figuratively speaking of course) so that they know what to expect next. Signposts such as this one are often useful to the readers especially within lengthy and complex dissertations.

- **Paragraph 6**. As anticipated, the paragraph offers plenty of examples of different types of signals used for different purposes. Note that new examples are being included here even though, arguably, it would be better to split the paragraph in two or more depending on the type of signal discussed (visual, chemical, tactile). Finally, note how the presence of a much needed hyphen in **food-sharing** makes the final sentence clearer than its previous version.

Even though further stylistic issues should be addressed, as I pointed out to the student, this revision represented a significant improvement on the previous draft. At least, the overall structure was in better shape and much of the content was there. However, not being familiar with her field of research, I was not in a position to advise the student on either the accuracy or the relevance of all the information presented and suggested to defer this aspect to her supervisor. Nevertheless, this example serves to show how even an nonexpert can offer useful feedback on the main issues with a draft and how these could be addressed.

Eventually, once the student discussed her progress with her supervisor, she sent me the following final version, the one appearing in her submitted PhD thesis.

CHAPTER ONE - GENERAL INTRODUCTION

(Submitted version)

1. Beaked whales have been the focus of more dedicated research in the last decade than ever before, following a number of mass stranding events co-incident with military naval exercises (Balcomb and Claridge, 2001; Van Bree and Kristensen, 1974; Cox et al., 2006; Evans and England, 2001; Fernández et al., 2005; Frantzis, 1998; Jepson et al., 2003; Simmonds and Lopez-Juraco, 1991). They are one of the largest mammalian groups (Dalebout et al., 2004), yet have been one of the least known (Wilson, 1992). This is in part because their behavioural characteristics make them difficult to study. They are typically found in small groups, are cryptic when at the surface, which is only ever for a brief period, and dive to great depths for extremely long durations making them difficult to detect (Barlow, 1999). This study focuses on Blainville's beaked whales, *Mesoplodon densirostris* (Blainville, 1817), also known as dense-beaked whales because of their dense skull structure.

continues...

continued...

2. Recent behavioural studies have focused primarily on beaked whales' responses to anthropogenic noise sources in order to quantify their movement away from potentially damaging sounds (Allen et al., 2014; DeRuiter et al., 2013; Tyack et al., 2011). This is in contrast to earlier studies on the effects of anthropogenic noise on other marine mammals, which primarily focused on their communication systems. For example, ship noise reduces the distances over which blue (*Balaenoptera musculus*) and fin (*Balaenoptera physalus*) whales are able to communicate (Payne and Webb, 1971), and causes North (*Eubalaena glacialis*) and South Atlantic right whales (*Eubalaena australis*) to alter their call frequency (Parks et al., 2007). Despite the recent interest in beaked whale biology, there is still no insight into beaked whale communication and the way it is affected by anthropogenic sounds. This lack of knowledge will inevitably hamper interpretation of behavioural response studies.

3. In this introductory chapter, I present an overview of the way in which different animal species communicate and discuss how their social structures influence their communication. A brief review on what is currently known about Blainville's beaked whales will also be presented, together with an outline of the thesis.

1.1 ANIMAL COMMUNICATION

4. Communication is ubiquitous in the animal kingdom. Animal communication occurs either by non-voluntary cues, or by signals, and is widely agreed to involve the transfer of information between a sender and a receiver that on average benefits both (Bradbury and Vehrencamp, 1998).

5. Cues are traits or behaviours that are obligate and inadvertent, and their emission does not generally benefit the producer, at least in terms of effects on other animals (Seeley, 1989); for example, a mosquito can detect the cue of CO_2 exhaled from a mammal upwind, thus gaining information on the mammal's location, which the mammal surely does not benefit from.

6. Yet some obligate cues can benefit the producer; for example, a male lion's (*Panthera leo*) mane provides a cue of his fitness to female lions (Schaller, 1972), and the urine of female Asian elephants (*Elephas maximus*) provides a cue of their sexual receptiveness to male elephants (Rasmussen et al., 1982). In the classical view of signal evolution, cues have been shown to evolve to become signals through ritualisation (Tinbergen, 1952), such as the canine behaviour of baring teeth. This cue, intimating the opponent to stay away, evolved from dogs baring their teeth immediately before biting, to get their lips out of the way (Krebs and Dawkins, 1984). The evolution of signals from cues is however limited by both physiological and ecological constraints (Arnold, 1992), which differ across species.

continues...

continued...

7. Signals, in contrast to cues, are deliberately given by a sender and have been selected because of the effect they have on altering the behaviour of a receiver (Wilson, 1975). Signals evolve in line with the response and performance of the receivers. Studies of mate choice in fish through the evolution of colourful signals have also shown the removal of these signals in some cases due to local ecological conditions. Male sticklebacks (*Gasterosteus* spp.) in lakes that do not have clear water have lost their bright red marking on their underparts, and the females have lost their preference for the red marking and their sensitivity to red light (Boughman, 2001).

8. Signals can vary in form, function and benefit to the signaller. Forms include visual, acoustic, tactile and chemical. Functions include reproductive success through sexual advertisement and mate attraction; alarm calls to alert conspecifics to the presence of a predator; conflict resolution; individual identification, used commonly in mother-offspring recognition; and maintaining group cohesion. Additionally, bats and cetaceans use acoustic signals to detect and localise prey.

9. Visual signals can be used to attract females and also to deter predators or competing males by providing an indication of fitness and dominance. The Puerto Rican crested anole (*Anolis cristatellus*) displays its ability to escape a predator by the number of push-ups it can perform during certain predation events. This display benefits the anole by also acting as a signal to potential mates on the individual's quality (Leal, 1999). Another type of visual display is seen in bowerbirds (family *Ptilonorhynchidae*), in which sexual advertisement has evolved away from their own plumage to the elaborate decoration of the bowers they build. This evolution has enabled bower birds to indicate through the bower quality, their age, experience, and dominance to the females, as in order to produce the most elaborate bower, they will have had to steal and possibly destroy another male's bower (Pruett-Jones and Pruett-Jones, 1994).

Now, for one last time, let us review again each paragraph in turn to see what changes have been made and how they have contributed to improving the overall readability.

- **Paragraph 1.** The paragraph now provides some more background information about beaked whales and explains why they have remained poorly known to date. The paragraph is well structured and entirely focussed on one key idea.
- **Paragraph 2.** This paragraph now provides a nice and concise overview of previous studies, where the distinction between studies which focussed

on beaked whales' *behaviour* and those which focussed on beaked whales' *communication* is now made very clear. In particular notice the adjective *behavioural* in **Recent behavioural studies**; the addition of *other* in **other marine mammals**; and the use of *primarily* in **primarily focussed on their communication system**. Also, the simple addition of **For example**, at the beginning of the third sentence, provides a better link with the previous sentence, making the purpose of the listed examples more obvious: to demonstrate the effect of sounds on other marine mammals.

The last two sentences now better accomplish what the student was trying to achieve in the previous version, namely to highlight the research gap and to provide a motivation for the study. Notice how the last sentence **This lack of knowledge will inevitably hamper interpretation of behavioural response studies** implicitly indicates a sense of urgency without the need for vague phrases such as **understanding their communication system is paramount** used in the previous version.

- **Paragraph 3**. This is now a nice transition paragraph that concisely signals to the reader what to expect next. Here, the use of the first personal pronoun is retained in the first sentence only, while the passive construction is used in the second. The combination of both makes for a varied style which is more pleasant to read and sounds less colloquial.

- **Paragraph 4**. Beautiful. In this paragraph, the student quickly presents a distinction between cues and signals and provides a reference to support the evidence that the exchange of information generally benefits both the sender and the receiver.

- **Paragraph 5**. Having briefly introduced the two main types of communication, the student now moves on to define in greater detail what cues are and provides a specific example in support. Notice how, in the example, the fact that the sender does not benefit from the cue is made more explicit than in the early draft by adding **which the mammal surely does not benefit from.**

- **Paragraph 6**. Here, notice how the **Yet** introduced at the beginning of the sentence immediately signals an exception about cues, namely that **some obligate cues *can* benefit the producer**. Also, notice how the example now has a clear parallel structure: **a male lion's mane provides a cue of his fitness...** and **the urine of female Asian elephants provides a cue of their sexual receptiveness**. The parallelism is made even more obvious by choosing to use the same words (**provides a cue**) for both examples.

- **Paragraph 7**. As done in paragraph 5 for *cues*, the student now provides more details about *signals*. She further states that **Signals evolve in line with the response and performance of the receivers**, a new concept that did not appear in previous versions. As a result, the examples provided are now different than those used previously, but all serve the purpose of supporting the statement.

- **Paragraph 8**. Here different types of signals and different functions are all briefly stated, before they are discussed in greater detail in following paragraphs.

- **Paragraph 9**. This paragraph entirely focuses on the first type of signals, namely visual signals, presented in the previous paragraph thus preserving a sense of order. The paragraph provides further details about the function of visual signals and supports each by specific and visually evoking examples. Again, each of these paragraphs is now well structured and mostly focussed on one key idea.

After reading this final version, I was extremely pleased to see how much the student had been able to improve on her earlier draft by following mine and her supervisor's feedback. I was however even more pleased when, soon after discussing her thesis with her examiners, she emailed me the following lines:

"I almost fell over when they commended me on my writing - so far as to say that one of the reasons they thought I should publish was to get that style of writing into the literature to encourage more people to be as clear!?!?"

As this book comes to a close, my hope for you is that you too may become a better writer and enjoy the long-lasting satisfaction that writing well will bring you.

III

Supporting Material

III

CHAPTER 9

Section Templates

CONTENTS

- 9.1 Introduction 144
  - 9.1.1 The purpose 144
  - 9.1.2 Building a template: A worked-out example 145
  - 9.1.3 Introduction: A template 147
- 9.2 Methods 148
  - 9.2.1 The purpose 149
  - 9.2.2 Methods: A template 150
- 9.3 Data analysis and results 151
  - 9.3.1 The purpose 151
  - 9.3.2 Data analysis and results: A template 152
- 9.4 Discussion and conclusions 152
  - 9.4.1 The purpose 152
  - 9.4.2 Discussion and conclusions: A template 154
- 9.5 Abstract 155
  - 9.5.1 The purpose 155
  - 9.5.2 Abstract: A template 155

INEXPERIENCED academic writers often struggle with knowing exactly what to write and where. Thus, it helps to have a model or template to follow when writing something for the first time. Building a template requires that we understand the intrinsic structure and purpose of each section or chapter. The best approach I found to discover the underlying structure of some text is that suggested by Hilary Glasman-Deal in her excellent book *Science Research Writing for Non-Native Speakers of English* [8]. As we shall see, the approach consists in focusing on the *function* of sentences rather than on their *content*.

Depending on your discipline, some of the section/chapter names used here may differ from those routinely adopted in your research area and you should familiarise yourself with any different conventions that may apply. Their purpose, however, will largely remain the same and with some practice you will be able to distinguish which section or chapter serves as introduction, methodology, data analysis and so on.

Following the approach presented in *Science Research Writing for Non-Native Speakers of English* [8], we shall first look at the *purpose* of each section, the *type* of information it should contain, and finally the *order* in which information is presented to the reader, so that it will be easier to uncover the underlying structure. The templates provided for each section (*Introduction, Methods, Data Analysis and Results, Discussion and Conclusions, Abstract*) are adapted from those presented in *Science Research Writing for Non-Native Speakers of English* [8] and will serve as a useful canvas for you to follow in your own writing.

Notice, however, that the process of writing is not linear and the order in which chapters and sections are presented to the reader may not reflect the order in which they are normally written. Indeed, as you will see, sections such as *Abstract* and *Introduction* are best written last!

9.1 INTRODUCTION

9.1.1 The purpose

The main purpose of an *Introduction* is to identify a *problem* or *gap* in current research. Depending on the nature of your project, the research gap may refer to some specific aspects that have never been addressed before, to contradictory results in the existing literature, to new lines of enquiry in your field, or to a combination of some or all of these.

Identifying a research gap should also naturally lead to a *research statement* that defines the aim of your study. Ideally, you should only undertake a study if:

a) you can address (and hopefully answer) a specific research question, and/or

b) your study can provide a valuable contribution to your field.

In fact, the existence of a research gap is not by itself a justification for undertaking a study. Especially important for research grants applications is that the introductions also explain and motivate *why* the research gap should be addressed in the first place and the research be funded.

In most scientific work, the *Introduction* is not the place for an extensive literature review. However, some key studies should briefly be reported to support the statements you make about the research gap and its relevance in a wider context. If in doubt about the right extension of your *Introduction*, consult the guidelines for authors in your target journal or read the introduction chapters of theses and dissertations within your research discipline.

Ideally, the *Introduction* should grab the readers' interest and make them want to find out more. This also requires that you spend some time to assess the likely background knowledge that your potential readers already possess.

If in doubt, or if you are addressing a mixed readership, err on the cautious side: write for the least knowledgeable of your readers and always strive to provide key information to help them understand what follows (Section 3.1).

Even though the *Introduction* appears as the first section (chapter) of your paper (thesis), it is probably best to write it last. This is because you need to know what results you have obtained in your study to be able to tell a coherent story and to state whether or not the original research question posed in your *Introduction* has been answered. Indeed, the research question is sometimes modelled around the specific results one has obtained in the course of the project and writing the *Introduction* last will allow you to be consistent.

9.1.2 Building a template: A worked-out example

Following the approach presented in *Science Research Writing for Non-Native Speakers of English* [8], you should now take a few minutes to read the following *Introduction* sample. For each sentence, identify the type of information presented and the specific *function* that each sentence fulfils. In other words, focus not on what the sentence actually says but rather on what the author is trying to achieve with each sentence and why (s)he put it there.

INTRODUCTION

1. Electronic Voting Systems (EVS), colloquially known as clickers, have become widely spread in Higher Education as a way to promote students' engagement within lectures and classes. **2.** EVS are very simple to use and allow instructors to pose a question on the screen and to invite students to vote. **3.** After polling is complete, a bar chart is displayed of the responses received.

4. EVS have been extensively used as diagnostic testing tools[1], for contingent teaching[2], and in peer instruction[3] and have found widespread application in disciplines such as biology[1,4], microbiology[5], physiology[6], and physics[3,7].

5. Yet, despite the general enthusiasm with which EVS are met by students and teachers alike, such technologies are by no means a "magic bullet" for effective learning. **6.** A key challenge resides in deploying EVS as part of a coherent pedagogy and in designing effective questions that promote deeper learning[8]. **7.** In particular, recent studies have demonstrated how students' engagement can be further promoted when learners are directly involved in the designing process.

8. This paper presents a case study on the use of EVS for a class competition where half of the Multiple Choice Questions used in the competition were generated by the students themselves. **9.** On the basis of the results obtained, we conclude that this approach only marginally enhances students' engagement over more traditional uses of EVS.

Once you have identified the function of each sentence, you can compare your answers to those given below.

> ## INTRODUCTION
>
> **Sentences 1-3** These sentences serve to establish the *importance* of the research topic and to provide some general background information. Key terminology is also succinctly explained or defined, if necessary, depending on the audience being addressed.
>
> **Sentences 4-7** These sentences offer a brief overview of some key projects in the research area, before providing a transition between the literature review and the general problem. A specific problem or gap in current research is also highlighted (in this specific example, this is done in sentence 6). This identifies the problem to be tackled in the present study.
>
> **Sentences 8-9** These sentences briefly describe the paper/chapter and announce the main findings of the study.

9.1.3 Introduction: A template

Based on the insight we have gained on the purpose of each sentence, we can now move to the next step, namely that of building a model, or template, for the *Introduction* section. To do so, we need to pay attention specifically to the order in which sentences are presented to the reader, the level of detail provided, and the way in which sentences are connected with one another, so as to guide the reader through a logical sequence. After reading a number of introduction sections, you will soon discover that they all have the same key elements, also likely presented in the same order.

A template for the *Introduction* is illustrated in Table 9.1. It consists of essentially three main components: an *opening*, a *development*, and a *closing* part. Clearly the length of each will depend on the level of details provided and that in turn depends on the type of document you are writing. For a research paper, each of these parts would typically consist in a couple of paragraphs; for a PhD thesis, each may be up to a couple of pages or so.

You can now use the template in Table 9.1 as a blueprint to follow when you write the *Introduction* section or chapter of your document and you should constantly refer to this template to make sure you include all the necessary information. With time, you will be able to write without any reference to the templates presented here as they will become a natural part of the way you write.

However, notice that, when used for the introduction to a research paper, the template presented here is only applicable to *Regular Articles*, namely those routinely submitted for publication as standard papers in a given journal. Introductory sections for articles that appear as *Letters* or *Rapid Com-*

Table 9.1 INTRODUCTION: A template
(Adapted from *Science Research Writing for Non-Native Speakers of English*, by Glasman-Deal [8])

Opening	briefly state the **importance** of your research area
	provide some **background** information
	recall **focus of current, on-going research**
Development	outline key recent research **contributions**
	highlight **research gap** (normally, the focus of your study)
	identify **specific problem(s)** to be addressed in your study
Closing	present an **overview** of the work done
	briefly mention **key results** obtained
	re-state aim of present work (typically, in a thesis)
	state directions for future studies (typically, in a paper)

munications or indeed those published on high-impact journals such as *Nature* or *Science* may differ considerably in structure and are typically shorter (a couple of paragraphs at most). For these types of publications, it is always good practice to follow the guidelines for authors provided by the journal in order to find out what information is specifically required in the introduction. Similar considerations apply to other sections as well.

Finally, remember that, while the model presented here is fairly typical for most scientific disciplines, you may notice some departures in your own field (for example, in highly technical, theoretical, or computational areas). Thus, it may be instructive at this point to read other *Introduction* sections within your own discipline and determine whether a similar pattern emerges. With some practice, you will soon be able to devise your own template, one that is specifically appropriate to your research field.

9.2 METHODS

9.2.1 The purpose

When we carry out experimental research, there is an implicit expectation that if we repeat the experiment under the same set of circumstances, the results we obtain should be the same within experimental uncertainties. This is perhaps the most distinguishing feature of the so-called *scientific method* and dates back to scientists as Nicolaus Copernicus (1473-1543), Galileo Galilei (1564-1642), and Francis Bacon (1561-1626) who are regarded as key founding figures of modern-day science. Conversely, the same experimental investigation carried out under different circumstances (e.g., different approach, equipment, or materials) may not - and indeed should not - lead to the same results.

The purpose of the *Methods* (or Methodology[1]) section is to provide your readers with *all* the necessary information to allow them to repeat your experiment and reproduce the same results. Now, you may wonder what is the point in repeating the same experiment and getting the same results? Why would scientists bother to do something that has already been done? The answer is simple: validation. Unless other researchers can independently verify the correctness of our results, there is arguably little point in doing research as nothing would withstand the test of time, nor would any discrepancy be discovered that could potentially lead to major new advancements.

In my experience, one of the greatest challenges students face is being able to separate the methodology from the data analysis. There seems to be a tendency to explain *how* something was done while at the same time saying *what* was obtained. This is often a cause for confusion for the readers.

If you also struggle with separating the method from the results, think of the following: the method or approach is something fairly generic that is independent of the specific experiment or study you are carrying out. For example, if I want to study a nuclear reaction and wish to identify the types of particles produced, I have a choice between different techniques: *time of flight*, ΔE-E *telescope, mass-over-charge ratio (A/q)*, and so on. Each of these approaches has its own advantages and disadvantages and I need to choose carefully the method that is best suited to the explicit purpose of my study. However, the approach itself will be independent of the specific reaction I intend to study and can easily be applied to other reactions as well. Thus, when describing the approach used, I can do so without any reference, for example, to the actual type of detectors used, their make and model, their thickness, their resolution, and so on. These details can be provided in full in a following chapter or section, describing the experimental setup.

---

[1] In some disciplines, the words *methods* and *methodology* have very distinct meanings: methods is used to indicate the tools, techniques or processes used in our research; by contrast, methodology refers to the study of *how* research is done, or - if you prefer - the justification for using a particular research method. Such a distinction is hardly relevant in the natural sciences and throughout this book these words will be used interchangeably to indicate the specific approach used in a research study.

Perhaps, the most important aspect of your choice of methods is making sure that the approach you select is fit for purpose. If it is not, then the validity of your study may be questioned and the results obtained, no matter how accurate or precise, may be completely invalidated.

Finally, you should use appropriate language to highlight that great care has been taken when carrying out specific procedures. Good examples of how this can be attained can be found in Glasman-Deal [8].

9.2.2 Methods: A template

Using the same approach highlighted in Section 9.1.2, you can now build a model for the *Methods* section (refer to [8] for a worked-out example). Once again, be aware of possible discipline-specific differences and make sure you familiarise yourself with the standard in your own research. A typical structure for the *Methods* section is summarised in Table 9.2, showing its key elements and their relative order.

Table 9.2 METHODS: A template
(Adapted from *Science Research Writing for Non-Native Speakers of English*, by Glasman-Deal [8])

Opening	briefly recall the **purpose** of your study
	provide a **general overview** of the approach used
	justify your choice of method
Development	provide information on **materials** and **equipment** used
	give **specific details** of makes, models, quantities, etc.
	use appropriate language to indicate that **care was taken**
	relate your materials/methods to **other, similar studies**
Closing	state any **problems** encountered
	recall any **limitations** of methods used

Notice that some elements in the template simply serve the purpose of reorientating the readers. This is especially important in long documents, such as a PhD thesis, where key aspects of the study (for example, the purpose of the work) might have been stated a few chapters earlier.

Some authors debate whether details such as make and model of a piece of equipment should be explicitly stated. The answer largely depends on the nature of the document you are writing. If you are writing a short article or a *Letter* you can avoid giving such details and point the readers to another, more extensive, paper or some online material where the approach and apparatuses used are fully described. On the other hand, if you are writing a PhD thesis, this is likely to be the only repository for recording all the details of the study you have undertaken. In this case, I would strongly advise you to be exhaustive and provide full details that would enable a reader to repeat your study if they so wish. Remember, reproducibility is the ultimate validation of any piece of scientific work. If the extent of such details proves too much for the main body of your work, consider placing them in an appendix.

9.3 DATA ANALYSIS AND RESULTS

9.3.1 The purpose

The purpose of the *Data Analysis and Results* (DAR) section (or chapter) is to provide the readers with an overview of the *key findings* of your study. Once again, the level of detail will depend on the nature of the document you are writing. However, even in the case of a PhD thesis, you should refrain from reporting on every single result obtained. Rather, you should focus on the most relevant results and present them clearly and concisely (refer to Chapter 7 for advice on how to prepare effective figures and tables).

It may also be worth explaining the overall approach used to analyse the data, especially if the analysis required lengthy and complex procedures before arriving at the final results. Bear in mind that the approach you used *to analyse* the data is different from the approach you used *to collect* them and it is often useful to separate these out for clarity.

Students sometimes ask me whether they should highlight problems encountered along the way both in terms of the approach used or the results obtained. The answer depends on the nature of the problem or the 'null' result. Occasionally, inconclusive results arise from poorly planned experiments and procedures. In such cases, these could be regarded as a mistake on the part of the researcher and need not be reported. Other times, however, inconclusive results are obtained despite flawless planning and execution. In such instances, it is always good practice to report on any 'dead-end roads' encountered. These can ultimately save precious time and effort for other scholars following in your footsteps!

Whatever results you are presenting, you should always state their statistical significance and detail the way in which errors have been determined. You should also use appropriate language to inform your readers about the way you perceive your results [8]. For example, if you obtain a given response say 30% of the times, you may present this as a strong result by saying **in as many as 30% of cases...**, or a weak one by saying **in only 30% of cases...**. This may seem in contrast with the objectivity required in scientific academic writing. However, it is important that you tell your readers what you think of your results so that they can follow your discussion and perceive your conclusion as a consistent story, even though they may disagree with your interpretation.

The *Data Analysis and Results* section (or chapter) is often written first, and for good reasons: one is that many students typically start writing up once they have completed their data analysis. As this will still be very fresh in their minds, the DAR chapter is an obvious one to write first. Another reason is that, having spent many months analysing your data, you know exactly what you did and how. Writing about your data analysis often proves easier to do than writing up, say, the conclusions. The DAR chapter can also provide a test bench for your writing and acquiring practice on something you are familiar with may prove less daunting. As we have seen in Chapters 5 and 6, just make sure you leave enough time at the end of the write-up process to allow for a thorough re-read that will allow you to iron out any inconsistencies in structure, content and style.

9.3.2 Data analysis and results: A template

The approach for creating a model for the *Data Analysis and Results* section is the same as we have already seen for the previous two sections. Here, we simply report the final template in Table 9.3. Be aware, however, that departures may be significant if your project is more theoretical or computational rather than experimental. If so, make sure you read enough samples of *Data Analysis and Results* sections (or their equivalent) in your discipline to find out the most common approach to structuring this section.

9.4 DISCUSSION AND CONCLUSIONS

9.4.1 The purpose

I am often asked about the difference between a *Discussion* and a *Conclusion* section (or chapter). The distinction is somewhat blurred and that is perhaps the reason why these two sections/chapters often come together under a single heading. However, I would argue that a *Discussion* is where you compare the results you have obtained in your study to those of researchers before you. In this sense, a key aspect of the *Discussion* section is to relate your study to previous research in order to put your findings in the right context. Thus, as the name suggests, this section would contain a discussion around possible reasons for any discrepancies between your results and previous ones; or even an explanation as to why your study supersedes previous findings, either yours

Table 9.3 DATA ANALYSIS AND RESULTS: A template
(Adapted from *Science Research Writing for Non-Native Speakers of English*, by Glasman-Deal [8])

Opening	recall the **research aims** of your study
	briefly re-state the **approach used**
	provide a **general overview** of the results obtained
Development	describe your **key results** in detail
	invite readers to view results in **Figures** and **Tables**
	discuss **(statistical) significance** of your results
	compare results to (model) **predictions**, if applicable
Closing	mention any **problems** encountered
	outline key **implications** of your results

or by others. The *Discussion* is also the right place to highlight any problems with the approach used in your study and whether, with hindsight, a different approach might be better suited to improve on current results.

By contrast, the *Conclusions* section would focus more on the *implications* that your results have in the wider context of your discipline. What is the greatest achievement of your study? What difference does it make in addressing the questions or research gap identified in the *Introduction*? In what way would your study influence any future research? Are there any limitations to its possible applications? These are the main questions that you need to address in the *Conclusions* section of your paper or thesis.

Some authors also use the *Conclusions* section as a way of summarising the entire study presented. This is certainly useful given that many scholars will read the *Conclusions* before the bulk of your paper (or thesis) just to assess whether your research is relevant to their project or not. It is therefore important that your *Conclusions* also fulfil the purpose of providing a concise overview of your entire work.

9.4.2 Discussion and conclusions: A template

Following the same procedure used in Section 9.1.2, you should now select a well-written *Discussion and Conclusions* section (or two distinct ones, if that is the case) from a regular article in your own discipline and identify the purpose of each sentence (or, for lengthy documents, of each paragraph). In particular, pay attention to the way in which information is recalled from previous sections (chapters) and observe how only the key findings and results are mentioned in these final sections. Ideally, you should be able to identify a pattern similar to that shown in Table 9.4 below. As always, be aware of possible discipline-specific differences. Try to find out which typical structure applies to your area of research and produce a template that can serve you as a guide when you write your own *Discussion and Conclusions*.

Table 9.4 DISCUSSION AND CONCLUSIONS: A template
(Adapted from *Science Research Writing for Non-Native Speakers of English*, by Glasman-Deal [8])

Opening	relate your results to **existing research**
Closing	highlight **key contribution(s)** of your study
	discuss **main implications** in broader context
Opening	revisit **previous sections/chapters**
	summarise **general** or **key results** obtained
Closing	discuss possible **limitations** and **applications**
	state directions for any **future work** (optional)

9.5 ABSTRACT

9.5.1 The purpose

The *Abstract* is the most read part of any paper and often the only one fully accessible through databases and online repositories. The abstract provides a brief, yet comprehensive, overview of the entire study so that other scholars can quickly decide whether this is relevant to their research and whether to invest time in reading the entire paper (or thesis).

In its most general form, a well-written abstract will report the main questions or problem (i.e., the purpose of the research) that the study aims to address, an outline of the methods used, a brief discussion of the results obtained, and some conclusions or implications. All of these within a few lines!

Because abstracts are fully open access and searchable from databases, non-standard symbols, abbreviations and acronyms should be avoided. Similarly, abstracts do not normally contain references, but if one is needed, the full bibliographical record should be provided directly in the abstract rather than at the end of the paper.

9.5.2 Abstract: A template

Depending on journals, the key elements of an abstract may be separated into individual headings as it is now a widely accepted practice, for example, for papers published on Physical Review C. For highly specialised journals, however, the readership is typically limited to scholars operating in the same discipline. In such cases, the abstract will normally contain only information on specific aspects of the work, typically the method, the results, and possibly their implications.

Thus, there are essentially two types of abstract models. One is similar to a summary and is very structured. It deals with the main sections of the research article/thesis. This type of abstract is typically used for conference proceedings, general interest seminars, or for wide-audience journal articles. The other type focusses primarily on one or two aspects of the study: typically the methods and the results, and is normally found in technical papers and in specialised journals. Both models are provided in the table below, with the second type being highlighted in grey.

As always, you should familiarise yourself with what is typical in your field. For PhD theses, refer to any specific guidelines provided by your own university as to what your thesis abstract should contain.

Table 9.5 ABSTRACT: A template
(Adapted from *Science Research Writing for Non-Native Speakers of English*, by Glasman-Deal [8])

Opening	briefly describe the wider **context** of your study
	present a summary of **current status**
	state **main aim** of your study
Development	describe **methodology/materials** used
	state **key results** and **contributions** of your study
	mention possible **implications** of your results
Closing	state **impact** in the wider context
	indicate limitations and future directions (optional)

Chapter 9: Section Templates
In a nutshell...

- In scientific writing, each section or chapter has a specific purpose and a clear, recognisable structure. By paying attention to how information is presented to the reader you can identify specific patterns for each of the sections (chapters) that make up a research paper (thesis) in your discipline.

- Templates for various sections (*Introduction, Methods, Data Analysis and Results, Discussion and Conclusions, Abstract*) are presented in this chapter.

- Non-standard articles, such as *Letters to the Editors* or *Rapid Communications*, may have structures that differ slightly from the ones proposed here. Similarly, different disciplines may follow slightly different structures. Always familiarise yourself with any subject-specific requirements of different journals and institutions.

EXERCISES

9.1 **Identify the structure** [30 minutes]. Read three *Introduction* sections from three different *Regular Articles* in your discipline. Identify the purpose of each sentence (or paragraph for lengthy sections). Hence, create a template for each of the three *Introductions*. Do the templates look similar? If not, in what respect do they differ?

9.2 **Identify the structure (again)** [30 minutes per section]. Repeat the exercise above for different sections and compare the templates you identify with those provided in this chapter. Can you find similarities? If not, how do your templates differ from the ones provided in this chapter?

9.3 **Create your own template** [20 minutes]. While most scientific writing follows very similar patterns and structures to those described in this chapter, some may be subject-specific. Create your own templates if different structures are routinely followed in your discipline.

9.4 **Over to you.** Use the templates provided in this chapter (or those you have created) to write the corresponding section.

CHAPTER 10

Elements of English Grammar

CONTENTS

10.1 Basic Terms and Definitions 160
    10.1.1 Clauses and sentences 160
    10.1.2 Subject 160
    10.1.3 Object 160
    10.1.4 Nouns 161
    10.1.5 Pronouns 162
    10.1.6 Adjectives and adverbs 162
    10.1.7 Prepositions and conjunctions 162
    10.1.8 Verbs 163
    10.1.9 Phrasal verbs 164
    10.1.10 Verb tenses 165
    10.1.11 Infinitives, participles, and gerunds 167
    10.1.12 Auxiliary and modal verbs 168
10.2 Similar meaning, different spelling 168
    10.2.1 *Due to* or *owing to*? 168
    10.2.2 *That* or *which*? 169
    10.2.3 *Fewer* or *less*? 169
    10.2.4 *Who* or *whom*? and other personal pronouns 170
10.3 Similar spelling, different meaning 170

G RAMMAR can be troublesome for both native and non-native speakers of English. So, having at least a basic understanding of key grammatical elements and rules will greatly help you to write well. This chapter presents some basic aspects of English grammar for quick reference. However, if English is not your mother tongue, make sure you always consult a good English grammar book or a native speaker, whenever in doubt.

10.1 BASIC TERMS AND DEFINITIONS

This section provides definitions for some key elements of English grammar. Additional resources are listed in the Further Reading section.

10.1.1 Clauses and sentences

A *clause* is a group of words that contains a verb (Section 10.1.8). A clause may be part of a *sentence* or may be a complete sentence in itself. Sentences are used to express a statement, a question, an exclamation, or a command and can either be *simple, compound,* or *complex.*

A *simple sentence* normally consists of one statement (namely, the *main clause*). For example:

> The samples were cooled with de-ionised water.

A *compound sentence* contains two or more clauses of equal status (both main clauses) linked together by a conjunction such as *and* or *but:*

The samples were cooled with de-ionised water and placed in air-tight containers.

A *complex sentence* contains two or more clauses. Of these, at least one is a main clause (that is, has complete meaning and can stand alone) and at least one is a *subordinate* clause, i.e., one that does not make sense on its own. For example:

The samples were cooled with de-ionised water and placed in air-tight containers (main clauses), **after being wrapped in aluminium foils** (subordinate clause).

10.1.2 Subject

Each clause (sentence) must at least contain a subject and a verb. The subject of a clause (sentence) is the person or thing that carries the action expressed by the verb. In all previous examples, the subject in the main clause is **The samples**. The subject of the subordinate clause in the last example is **air-tight containers**.

10.1.3 Object

The grammatical object of a sentence is the person or thing that receives the action expressed by the verb. Some sentences, like the ones above, do not have an object; others may have more than one. The object can be a simple noun, as in:

> We (subject) **cooled the samples** (object) **with de-ionised water**

or a whole clause:

We noticed (main clause) that the samples' temperature had dropped by 10 °C (objective clause).

10.1.4 Nouns

A noun is a word representing a thing or a person. Nouns can be grouped into the following categories:

- *Common*: everyday words that represent some *concrete* or *abstract* (see below) object, feeling, place, idea, etc.
- *Proper*: names of a place, a person, or an institution. They are always capitalised, as in India, Elizabeth, The University of Edinburgh.
- *Countable*: common nouns that can take a plural. As such, they can combine with numerals or quantifiers and can take an indefinite article *a* or *an*: a book, an orange, three dosimeters.
- *Uncountable*: common nouns that cannot take a plural. As such, they cannot combine with numerals or quantifiers and cannot take an indefinite article; for example: furniture, luggage. These words are used with singular verbs.
- *Collective*: nouns referring to groups of individuals or entities. Examples include: committee, family, police. In English, these nouns can be used with a singular verb when referring to the entity as a group, or with a plural verb when referring to the individual members. For example:

    A committee (group) was appointed to short list the applicants

    but

    The committee (individual members) were impressed by the candidate.
- *Concrete*: nouns that refer to physical entities and can typically be perceived by one or more of our senses. For example: dog, toothbrush, detector, thermometer.
- *Abstract*: nouns representing ideas, qualities, feelings, concepts or other intangible objects. Examples include: theory, research, ability. Occasionally, the distinction between concrete and abstract nouns is not clear cut. For example, music can either identify an abstract concept as well as a specific musical piece. In scientific writing, many abstract nouns (typically those ending in -(t)ion, -ment, -ance, -ness, -ism, -ty) are often used in lieu of a verb. Because abstract nouns require a little more brain processing than concrete nouns or verbs, it is generally better to use the corresponding verb instead, as suggested in Section 5.3.2.

10.1.5 Pronouns

Nouns can sometimes be replaced by *pronouns* (literally something standing in place of a noun) to avoid repetition. For example:

We carefully mounted the supporting frames and secured *them* with screws.

Pronouns can be personal (he, you, they, them, who, ...) or impersonal (it, one, that, which, ...). Note that the personal pronoun you may sometimes be used impersonally, as in:

You can't really find anything in this mess (meaning one can't).

Pronouns must concord with the nouns they replace and it must be absolutely clear what they refer to, so as to avoid misunderstanding or ambiguity. For example, in the sentence ([27], p. 19):

The cells divided in the modified medium and formed clumps that were visible with the naked eye. This showed...

it is unclear what This refers to: is it the fact that cells divided in the modified medium, that the cells formed clumps, or that the clumps were visible to the naked eye?

In all cases where ambiguity can be an issue, consider rephrasing. This is especially important in scientific writing where accuracy of language is crucial.

10.1.6 Adjectives and adverbs

Adjectives and *adverbs* are often confused and most people do not appreciate the difference between them. Their most basic distinction stems from the way they work: adjectives qualify (or describe) a *noun* or a *pronoun*; adverbs qualify (or describe) an *adjective*, a *verb* or another *adverb*. Adverbs normally answer questions about *how, when, where, why*, or *to what extent, how often* or *how much* (e.g., daily, completely).

> He gave a quick (adjective) reply
> He replied quickly (adverb)
> He gave a very (adjective) quick (adjective) reply
> He replied very (adjective) quickly (adverb)

10.1.7 Prepositions and conjunctions

Prepositions are short words used to connect nouns, pronouns, or phrases to other words in a sentence. They are normally placed directly in front of nouns or, in some cases, in front of verb forms ending in *-ing* (called *gerunds*). Examples of prepositions are: in, on, at, to, into, for, with, after and so on. In academic writing, you should ideally avoid overusing prepositions as they may compromise the clarity of a sentence.

Conjunctions serve to link clauses together. Examples include: and, or, but, however, nevertheless, therefore, etc.

10.1.8 Verbs

Verbs are a key component of any sentence as they express the action that conveys meaning to the sentence. Without a verb, a sentence would simply make no sense.

Verbs can be *transitive* or *intransitive* depending on whether they can hold a direct object or not.

- **Transitive verbs** can be used either in the *active* or *passive voice*, depending on whether the grammatical subject of the sentence is also the one performing the action expressed by the verb, as in:

  We analysed the data (active voice)

  or whether the grammatical subject of the sentence is the one receiving the action expressed by the verb, as in:

  The data were analysed by us (passive voice)

  Note that the agent (i.e., the person or thing that performs the action) is always preceded by *by* in the passive construction. The agent may be omitted when unknown, irrelevant, or intended to remain undisclosed, as in:

  A mistake was made when venting the chamber and all foils raptured.

  Since the active voice is the one we normally use in the spoken language – after all, we say: I ate an apple for lunch today rather than An apple was eaten by me for lunch today – our brains understand active sentences more quickly than passive ones. For this reason, you should ideally try to use the active voice whenever possible. Note, however, using the active voice does not necessarily imply having to use personal nouns or pronouns. For example:

  Paracetamol is shown to act on the nervous system (passive voice)

  can easily be rephrased in the active construction as:

  Paracetamol acts on the nervous system (active voice)

Indeed, when used at all costs, the passive voice may also end up in rather cumbersome and stylistically inelegant sentences as, for example:

The origin of the huge variety of elements is tried to be answered within the field of nuclear astrophysics.

Far better is to rephrase the sentence instead as:

Nuclear astrophysics aims to investigate the origin of chemical elements (in the universe).

- **Intransitive verbs** cannot hold an object and as such cannot be used in the passive form. An example is: **I exist**.

  Note, however, that some verbs (for example **to live**) can have both a transitive and intransitive use as in: **I live in Milan** (intransitive use), or: **I have lived a wonderful life** (transitive use).

Finally, some sentences do not have a proper subject and are called *impersonal*, as in **It is raining**.

10.1.9 Phrasal verbs

Many English verbs (called *phrasal*, or *prepositional verbs*) can combine with a preposition to form a single semantic unit: **write down, look after, run into** are all examples of phrasal verbs. Like any other verb, phrasal verbs can also be transitive or intransitive (Section 10.1.8).

Ideally, you should try to avoid using phrasal verbs and use more formal equivalents. So, rather than saying:

All possible sources of systematic uncertainties were carefully looked at

use instead:

All possible sources of systematic uncertainties were carefully examined

However, if you do use a phrasal verb, keep the preposition close to the verb even if at the end of a sentence[1], as in:

For the purpose of this application, the host institution is defined as the place where most of the research will be carried out.

---

[1] In formal writing you should normally avoid ending a sentence with a preposition. The only case where this is allowed is if the preposition forms part of a phrasal verb.

Table 10.1 Verb tenses (present, past, and future) conjugated in their various forms (simple, continuous, perfect, and perfect continuous).

	Present	Past	Future
Simple	The cells grow	The cells grew	The cells will grow
Continuous	The cells are growing	The cells were growing	The cells will be growing
Perfect	The cells have grown	The cells had grown	The cells will have grown
Perfect continuous	The cells have been growing	The cells had been growing	The cells will have been growing

10.1.10 Verb tenses

Verb tenses serve to indicate *when* something happened. In English, there are three main tenses: *present, past* and *future*. Each can be conjugated in four different forms, *simple, continuous, perfect* and *perfect continuous* as summarised in Table 10.1.

- **Present tenses**

  The *present simple* is normally used to indicate habitual actions (**I get up** at 7am every morning); universal truths (**the Sun rises in the East**); results of accepted validity (**the DNA has a double helix structure**); and definite arrangements in the future (**I will phone you when I get home**).

  The *present continuous* refers either to events happening right now (though not necessarily at the precise moment of writing/reading) or to arrangements that will take place in the near future.

  The *present perfect* is used for events that happened in the near past at some unspecified time; while the *present perfect continuous* refers to something that started in the past and is still happening.

- **Past tenses**

  The *past simple* refers to things that happened in the past (with or without an explicit time mentioned).

  The *past continuous* expresses some past action that continues over a period of time. It is often used with the past simple to indicate that something was happening before a certain point in the past, as in: The experiment *was working* (past continuous) until the lightening *stroke* (past simple).

The *past perfect* is used as a past equivalent of the present perfect when a time in the past is mentioned (The solution *had evaporated* completely by 2pm).

The *past perfect continuous* is used to establish a time relationship with some other events in the past (The detectors *had been working* for a few hours by the time we decided to switch them off).

- **Future tenses**
  The *future simple* is used to refer to events that will occur in the future or that we believe will happen.

  The *future continuous* expresses a continuous action that will occur in the future before another point in time (The turbo pump *will be working* at full speed by the time we arrive at the lab).

  The *future perfect* is used for events/actions that we *believe* will take place by a certain point in the future The cells *will have grown* by the time we get to the lab.

  Similarly, for the *future perfect continuous* (The cells *will have been growing* by the time we get to the lab), which is however less frequently used and may imply that the same action will continue after a certain point in the future, unlike the *future perfect*. For example, By December, I *will have been working* on this thesis for three months may imply you will still continue working on it after December. By contrast: By December, I *will have worked* on this thesis for three months may imply you will not continue working on it after December.

In general, you should avoid shifting tenses within the same unit of text, unless for a very specific reason. Consider, for example, the following sentences:

Although the phenomenon was fully investigated in the past, little attention *was paid* to its possible applications to other areas.

and

Although the phenomenon was fully investigated in the past, little attention *has been paid* to its possible applications to other areas.

In the first sentence, both clauses maintain the same past simple tense (was investigated and was paid). This implicitly implies that the search for possible applications was not of interest in the past and is likely not of interest now.

By contrast, in the second sentence, the past simple tense (**was investigated**) in the first clause switches to the present perfect (**has been paid**) in the second clause. This signals to the reader that while the search for possible applications was not of interest in the past, it may be of interest now.

A more in-depth discussion on the appropriate uses of verb tenses, together with illustrative examples, can be found in *Science Research Writing For Non-Native Speakers of English* by Glasman-Deal [8].

10.1.11 Infinitives, participles, and gerunds

- **Infinitives**

  An *infinitive* is a verb in the form *to-*, for example **to stand**. In academic writing, one should normally avoid splitting an infinitive by putting an adverb between the *to-* and the verb (**to proudly stand**) as some examiners/reviewers regard this as bad English. On the other hand, reformulating sentences to avoid split infinitives can sometimes make the sentence sound clumsy or strange. In these cases it is acceptable to use a split infinitive.

- **Participles**

  Participles come in two forms: present and past.

  Present participles always end in *-ing*, as in **doing, making**.

  Past participles end either in *-ed*, as in **formed, marked, stated**; or in other irregular forms, as in **bought, sung, spent**[2].

  Participles are used to form composite tenses of verbs as in Table 10.1. They can also be used as adjectives: all *broken* valves were discarded; the *dangling* cable was secured to the bar; or nouns: *critiquing* the work of others is normal practice in academia.

  When using a present participle, always make sure that it clearly refers to a specific noun and avoid a *dangling participle*[3]. An example of a dangling participle is given below:

  The samples were shown to have deteriorated *using* a microscope.

  Clearly, the samples cannot use a microscope and so the present participle *using* does not refer to any noun in the sentence. Rephrasing the sentence as:

  Using a microscope, we noticed that the samples had deteriorated

  eliminates ambiguity and makes it clear who does the act of **using**.

---

[2]Verbs forming the past participle in *-ed* are called *regular*; the others *irregular*. Always consult a dictionary if you are unsure about the past participle of an irregular verb.

[3]A participle is said to be *dangling* when used to modify a noun that does not actually appear in the text.

- **Gerunds**
  Occasionally, a present participle can be used as a noun, as in:

  Walking is good for you.

  In such instances, the present participle is called a *gerund*.

10.1.12 Auxiliary and modal verbs

- **Auxiliary verbs** are verbs that add functional or grammatical meaning to the main verb and are used to form tenses or to express modality, voice, or emphasis. The main auxiliary verbs in English are **to be, to have** and **to do**. Normally, the verb *to be* is used to construct the passive voice, the verb *to have* to construct various composite tenses, and the verb *to do* to form negative statements, questions, or to provide emphasis as, for example, in: I *do* see you.

- **Modal (auxiliary) verbs** are **can, could, may, might, must, ought to, shall, should, will** and **would**. These are used to express necessity, possibility, intention or ability to various degrees of certainty. It is important you familiarise yourself with their meaning and usage especially if you are not an English native speaker. For an excellent overview on the use of such modal verbs, see [8] or consult a good grammar.

10.2 SIMILAR MEANING, DIFFERENT SPELLING

Many words in English have a very different spelling but the same meaning. However, because their grammatical function is slightly different, they should not be used interchangeably. Most common examples are discussed below.

10.2.1 *Due to* or *owing to*?

Another common mistake concerns the improper use of **due to** and **owing to**. Again the meaning is the same, but because their grammatical function is different, they cannot be used interchangeably. Perhaps the easiest way to help you decide which one to use is by replacing them, respectively, with *caused by* and *because of*, as in:

The experiment failed owing to (meaning *because of*) poor planning

but

The experiment failure was due to (meaning *caused by*) poor planning

Note that **due to** is always used with the verb *to be* in any of its various forms (*is, are, was, were, been*).

10.2.2 *That* or *which*?

The pronouns *that* and *which* have the same meaning and can both be used to introduce *relative clauses*. Although they are often used interchangeably, the meaning of a sentence can change significantly, as illustrated by the following examples taken from a thesis on the study of the behaviour of beaked whales observed at sea. Consider the sentence:

> We have run an association matrix for the 15 adult females in the population, WHICH were seen more than 10 times

Here, the clause introduced by *which* is a *non-restrictive* one: it merely provides additional information and it can be removed without altering the meaning of the sentence, as also indicated by the comma that precedes *which*. In this case, the meaning of the sentence is: We have run the association matrix for the 15 adult females in the population. Incidentally, these adult females were seen more than 10 times.

By contrast, in the sentence:

> We have run an association matrix for the 15 adult females in the population THAT were seen more than 10 times

means: we run the association matrix *only* for those individuals that were seen more than 10 times. In other words, the clause introduced by *that* defines a sub-group of the population: "that one" and not another. The clause is therefore a *defining* or *restrictive* one, i.e., it cannot be removed without also altering the sense of the sentence. Note that a restrictive clause could also be used with *which* provided no comma is placed before it, as in:

> We have run an association matrix for the 15 adult females in the population WHICH were seen more than 10 times

So, in summary, both *which* and *that* introduce relative clauses. The first can be used for both restrictive and non-restrictive clauses (this latter without a comma), while the second can only be used with restrictive clauses.

10.2.3 *Fewer* or *less*?

Use **fewer** with countable nouns, as in:

> Our fitting procedure used fewer parameters than previous studies.

Use **less** with non-countable nouns, as in:

> Setting up the experiment required less work than anticipated.

10.2.4 *Who* or *whom*? and other personal pronouns

All personal pronouns except *you* (I, me, he, him, she, her, we, us, they, them who, whom, whose) change form depending on their function in the sentence, i.e., whether they are used as subject or object. So, for example:

> Who saw you at the party?

Here, *who* is the subject doing the act of seeing and *you* is the object. Compare with:

> Whom *did you* see at the party?

In this case, the person performing the act of seeing is *you* and *whom* becomes the object (the person being seen). Similarly, **I** and **we** are used as subject, while **me** and **us** as object. So, for example:

> John and I saw *you* at the party

but

> Did you see *John* and *me* at the party?

Finally, note that **whose** means *of whom* and should not be confused with **who's** which means *who is*.

10.3 SIMILAR SPELLING, DIFFERENT MEANING

Many words in English share very similar spellings but have completely different meanings. Table 10.2 provides a list of some words commonly used in scientific writing that can be easily confused.

Table 10.2 Commonly confused words often found in scientific writing.

accept	to agree to receive or do
except	not including
adverse	unfavourable; harmful
averse	strongly disliking; opposed
advise	(verb) to recommend
advice	(noun) recommendation
affect	(verb) to influence, to have an impact on
effect	(noun) a result, a consequence; (verb) to bring about, to cause
all together	all in one place; all at once
altogether	completely; on the whole
along	moving or extending horizontally on
a long	referring to something of great length
artefact	artificial effect, often due to experimental error
artifact	object produced or shaped by human craft
bad	not good
badly	(adverb) not well; in a bad manner; incorrectly
bare	naked; not covered
bear	(verb) to carry; to put up with
born	having started life
borne	carried
complement	to add to so as to improve; an improving addition
compliment	to praise or express approval; (noun) an admiring remark
discreet	careful not to attract attention
discrete	separate and distinct
ensure	to make certain that something will happen
insure	to provide compensation for death or loss
estimation	the process of finding an estimate
estimate	approximate calculation or evaluation
imply	to suggest indirectly
infer	to draw a conclusion
loose	to unfasten; to set free; (adjective) unfastened, large
lose	to be deprived of; to be unable to find
practise	to do something regularly
practice	use of an idea/method; business of a doctor, dentist, etc.
principal	most important; the head of a school
principle	a fundamental rule or belief

Chapter 10: Elements of English Grammar
In a nutshell...

- Verbs express the action that gives meaning to a sentence. Other key parts, which may or may not be appear in a sentence, include a subject (the person or thing that carries the action expressed by the verb) and possibly an object (the person or thing that receives the action).

- *Independent clauses* can stand alone; *dependent clauses* require another one to make full sense.

- Verb tenses identify the time at which an action took place. Normally, procedures described in methods sections are expressed with past tenses; results (especially if generally accepted to be true) are typically expressed in the present tense.

- Modal verbs (*can, could, may, might, will, would, should*) express different degrees of certainty and/or probability.

- Some words often used in scientific disciplines have similar spellings but very different meanings. Familiarise yourself with the meaning of each. If in doubt, consult a dictionary.

FURTHER READING

Grammar A-Z
  https://en.oxforddictionaries.com/grammar/grammar-a-z

Common Grammar Mistakes
  http://en.oxforddictionaries.com/usage/commonly-confused-words
  https://en.oxforddictionaries.com/grammar/the-comma-splice
  https://https://www.copyblogger.com/grammar-goofs/
  https://blog.hubspot.com/marketing/common-grammar-mistakes-list
  http://marialuisaaliotta.wordpress.com/2012/04/28/its-its-isnt-it/

Top 20 Mistakes in Undergraduate Writing
  https://undergrad.stanford.edu/tutoring-support/hume-center/resources/student-resources/grammar-resources/writers/top-twenty-errors-undergraduate-writing

Verb Tenses
  https://en.oxforddictionaries.com/grammar/verb-tenses

Mistakes with Passive Voice
  http://www.quickanddirtytips.com/education/grammar/avoid-this-common-passive-voice-mistake

References

[1] Tice, D.M. and Baumeister, R.F. (1997). *Longitudinal study of procrastination, performance, stress, and health: The costs and benefits of dawdling.* Psychological Science 8.6, 454-458, cited in https://motivationgrid.com/4-main-causes-procrastination-revealed/

[2] Goodson, P. (2013). *Becoming an Academic Writer – 50 Exercises for Paced, Productive, and Powerful Writing*, SAGE Publishing Ltd.

[3] http://pomodorotechnique.com

[4] http://christinekane.com

[5] Lamport, L. (1994). *LaTeX: A Document Preparation System*, Addison-Wesley Professional (second edition).

[6] Mittelbach, F., Goossens, M., Braams, J., Carlisle, D., Rowley, C. (2004). *The LaTeX Companion*, Addison-Wesley Professional (second edition).

[7] http://bibtex.org

[8] Glasman-Deal, H. (2010). *Science Research Writing for Non-Native Speakers of English*, Copyright © 2010 by Imperial College Press.

[9] Ridley, D. (2012). *The Literature Review. A step-by-step guide for students*, SAGE Publications Ltd. (second edition).

[10] Machi, L.A. and McEvoy, B.T. (2009). *The Literature Review: Six Steps to Success*, SAGE Publications Ltd.

[11] Aldridge, J. and Derrington, A.M. (2010). *The Research Funding Toolkit*, SAGE Publications Ltd.

[12] Crawley, G.M. and O'Sullivan, E. (2016). *The Grant Writer's Handbook. How to write a Research Proposal and Succeed*, Imperial College Press.

[13] http://sethgodin.typepad.com/seths_blog/2011/09/talkers-block.html

[14] Greene, A.E. (2013). *Writing Science in Plain English*, The University of Chicago Press.

[15] Gray, T. (2005). *Publish & Flourish: Become a Prolific Scholar*. Springfield, IL: Teaching Academy, New Mexico State University (first edition).

[16] Gray, T. (2015). *Publish & Flourish: Become a Prolific Scholar*. Las Cruces, NM: New Mexico State University Teaching Academy (second edition).

[17] Caenepeel, M. (2012). *Effective Scientific Writing*, Marie Curie Doctoral Training – Course Handbook, University of Edinburgh (unpublished).

[18] Murray, N. and Hughes, G. (2008). *Writing Up Your University Assignments and Research Projects*, Open University Press.

[19] Sword, H. (2016). *The Writer's Diet*. The University of Chicago Press.

[20] http://writersdiet.com/test.php

[21] Griffies, S.M., Perrie, W.A. and Hull, G. (2013). *Elements of Style for Writing Scientific Journal Articles*, Elsevier.

[22] Strunk, W. Jr. and White, E.B. (1999). *The Elements of Style*, Pearson (fourth edition).

[23] Zinsser, W. (1976). *On Writing Well*, HarperCollins Publishers (seventh edition, 2006).

[24] The American Institute of Physics (1990). *AIP Style Manual*
Available for download at:
http://web.mit.edu/me-ugoffice/communication/aip_style_4thed.pdf

[25] Thompson, A., and Taylor, B.N. (2008). *Guide for the Use of the International System of Units (SI)*, NIST Special Publication 811, 2008 Edition
Available for download at:
http://ws680.nist.gov/publication/get_pdf.cfm?pub_id=200349

[26] Hughes, I.G. and Hase, T.P.A. (2010). *Measurements and their Uncertainties. A practical Guide to Modern Error Analysis*, Oxford University Press.

[27] Koerner, A.M. (2008). *Guide to Publishing a Scientific Paper*, Routledge.

Index

A
Abstract
    purpose of, 155
    template, 155, 156
Abstract nouns, 74–75
Academic writing
    accuracy, 70
    cautiousness, 70, 71
    clarity, 4, 67, 70, 71 (*see also* revising)
    conciseness, 4, 70–71
    craft of, 5
    defining, 3
    objectivity, 70, 71, 72
    overview, 69–70
    PhD process, when to start writing during, 6
    pleasantness of, 4
    process of, 5
    scientific disciplines, in, 3, 70
    specificity, 70
    weekly checkup (*see* weekly writing checkup)
Achievements, recognizing, 12
Acknowledgments, 111
Acronyms, listing of, 110, 111
Action verbs, 73, 74
Active voice, 75–76, 81
Appendices, 109–110, 111
Audience
    defining, 39–40, 48
    PhD thesis, for, 40, 41
    research grant proposal, for, 41
    research paper, for, 41

B
BibTeX, 20

C
Cayley, Rachel, 59
Chronological structure, 57
Citations, 93–94
Compare and contrast structure, 57
Core dump, 47–48
Core sentences, 61

D
Dangling participles, 53, 86–87
Data analysis and results (DAR)
    overview, 151–152
    template, 152
Deductive reasoning, 3
Discussion and conclusions section
    purpose of, 152–153
    template, 154
Distractions, managing, 9, 15

E
Editing, 81. *See also* proofreading
    awkward writing, 78
    defining, 67, 81
    lean writing, 77–78
    negative statements, 79–80
    redundancy, 79
    scientific style, for, 72, 81
    spell checks, 92–93
    wasted words, 78
Elements of Style, The, 72
Equations, 108
EThOS, 27
Examples
    Blainville's beaked whales, 127–128, 129–130, 131–132, 133–134, 135–136, 136–138, 138–140

colorectal cancer, 120–123,
123–126, 126–127
energy consumption in data
centers, 115, 116–118,
118–119
Experimental results, reporting, 109

F
Figures, 102–103
captions, 103–104
preparations, 104–106
Formatting, 83, 97
Framing your time, 8, 9, 15

G
Glasman-Deal, Hilary, 24, 143
Glossaries, 110
Godin, Seth, 41
Goodson, Patricia, 7, 52
Google Scholar, 27
Grammar, 56, 65, 172
adjectives, 162
adverbs, 162
ambiguity, 87–88
clauses, 160
common errors, 84, 85, 86, 87–88
complex sentence, 160
compound sentence, 160
conjunctions, 163
dangling participles (*see*
dangling participles)
gerunds, 168
infinitives, 167
nouns, 161
objects, 160–161
overview, 159
participles, 167
prepositions, 162–163
pronouns, 162
similar meaning, different
spelling, 168, 169, 170
similar spelling, different
meaning, 170, 171*t*
simple sentence, 160
singular/plural forms, 85, 86

subject-verb agreement, 85
subjects, 160
subordinate clause, 160
verb tenses, 165–167, 172
verbs, 163–165
verbs, auxiliary, 168
verbs, modal, 168, 172
Greene, Anne, 56, 67
Grey literature, 27

H
Harvard system, 94, 95, 96
Homophones, 85

I
Independent clauses, 90
Information
gathering/organising, 20
reference management (*see*
references)
sources of, 19
InSPIRE, 27
Introductions, 144–145, 146, 147–148

J
Journal clubs, 27, 28, 29, 30
Journals
clubs for (*see* journal clubs)
high-impact, 41
specialised, 41
submitting to, 41

K
Kane, Christine, 9, 10, 11

L
LaTeX, 20
Layout, structured, 45–46, 48
Literature reviews
critical assessment, 27–28
databases, 27
defining, 26
example of, 35
functions of, 31
importance of, 32

location within the work, 31–32
matrix, 33–34, 36
meta-analysis, 31
meta-synthesis, 31
organizing, 33
process, 26–27
systematic, 31
traditional/narrative, 31
writing, 32
Logical sequence, 57

M
MEDLINE, 27
Methods section
purpose, 149–150
template, 150, 151
Mind maps, 43–46, 48
Mindset of a writer, 6–7
Multitasking, 9

N
National Institute for Standards and Technology, 108
Note-taking while reading, 21, 22. *See also* Paper Annotation Tool

O
Outlining, reverse, 59–60
Oxford comma, 89

P
Paper Annotation Tool, 21, 22, 23–24, 27, 36
Paragraphs as building blocks, 58–59
Parallel structure, 61–63
Passive voice, 75–76
Personal pronouns, 76–77
Phrasal verbs, 164–165
Pomodoro Technique, 8
Primary literature, 94
Problem to solution structure, 57
Procrastination
causes, 7
defining, 7
overcoming, 7, 8
Productivity, 9
Proofreading, 97
checklist, 96
grammar errors (*see* grammar)
importance, 83, 97
spell checks, 92–93
typos, 84
Punctuation
apostrophes, 91–92
colons, 90
commas, 88–89, 90
hyphens, 91
misuses of, 88
semi-colons, 90
Pyramidal structure, 57

Q
Quantities, 108

R
References
formats, 94–96, 97
management of, 20
software for, 20
Register, 3, 55
Reverse outlining, 59–60
Revising
clarity, 53–55, 65
feedback, 63–65
grammar, 56, 65
language, 55–56, 65
organization, 51, 52–53, 65
referencing, 53
structure, 56–58

S
Scopus, 27
Secondary literature, 94
Self-assessment, writing skills, 12, 13, 14
Skills, writing, self-assessment. *See* self-assessment, writing skills
Strunk and White, 72

Style, choosing and following, 25–26, 72–73
    abstract nouns, 74–75, 79
    action verbs, 73, 74
    passive *versus* active voice, 75–76
    personal pronouns, 76–77
    scientific, 69–72
Sunday Summit Form, 9
Sword, Helen, 68, 73
Symbols, 108

T
Table of contents, 101–102
Tables, 102
    footnotes, 107
    titles, 107–108
    usefulness, 107
Templates for structure, 24
Time management, 8
Time, framing, 8, 9, 15
Titles, 99–101
Tone, 3
Transition sentences, 60
Transition words, 78–79
Transitive verbs, 163–164

U
Units, 108
Uplevel You, 10

V
Vancouver system, 94
Variables, 108

W
Web of Knowledge, 27
Weekly writing checkup, 9, 10, 11
Writer's block, 42
Writer's Diet Test, The, 68–69, 73, 81
Writing skills self-assessment. *See* self-assessment, writing skills

Z
Zinsser, William, 72